QP
321
H5.9

1/71

D0163501

FIRST AND LAST EXPERIMENTS IN MUSCLE MECHANICS

A. V. HILL

CAMBRIDGE

AT THE UNIVERSITY PRESS

1970

RAMSEY LIBRARY
UNC ASHEVILLE
ASHEVILLE, N.C

Published by the Syndics of the Cambridge University Press
Bentley House, 200 Euston Road, London N.W.1
American Branch: 32 East 57th Street, New York, N.Y.10022

© Cambridge University Press 1970

Library of Congress Catalogue Card Number: 73–96092

Standard Book Number: 521 07664 1

Printed in Great Britain
at the University Printing House, Cambridge
(Brooke Crutchley, University Printer)

D. HIDEN RAMSEY LIBRARY
U. N. C. AT ASHEVILLE
ASHEVILLE, N. C.

To
RYOTARO AZUMA
physiologist, oarsman,
public servant,
and friend

CONTENTS

CONTENTS

PREFACE

In 1964 my last papers on the heat production of muscle were published: and about time too. After that I returned to another favourite theme, one that came to mind in 1920 but started in earnest with Herbert Gasser in 1923. It turned up at intervals later, particularly in a paper of 1949 called *The abrupt transition from rest to activity in muscle*. That paper triggered off a general interest in the growth and decay of the active state after a stimulus. But I have long realized that the methods and ideas of 1923 and 1949 had obvious defects and that the transition was not as 'abrupt' as I claimed. So in 1964 I set out to do the experiments better.

This took a long time, some of it spent in improving methods, more of it wasted in repeating elegant experiments which persisted in giving the same result. There was something missing, which appeared only after the experiments had changed course: in which there may be a moral.

But a number of interesting things turned up; some were new and exciting, some answered questions already asked, others posed new questions without decisive answers. Finally, early in 1967, it was decided that we should return to Cambridge (after 47 years); which was quite a job and meant abandoning a laboratory in London and all the instruments accumulated since 1920. In Samuel Johnson's words, 'When a man knows he is to be hanged in a fortnight, it concentrates his mind wonderfully.' In fact it was five months, not a fortnight, and returning to Cambridge was not a bit like being hanged; but the stimulating effect on my mind was the same, and some of the most interesting things turned up during that rather hectic period.

All of which left a vast amount of experimental material requiring critical analysis. The results, and the conclusions to be drawn, were too variegated to be put into ordinary scientific papers. Besides, it would be better to speculate rather at large, to 'let oneself go' in a way that editors of scientific journals quite properly forbid. That would require a book; and I recalled that in 1932 the Cambridge

University Press published a little volume of mine on *Chemical Wave Transmission in Nerve*. The Press was ready to consider a second one; the interval of thirty-five years gave reasonable assurance that it would be the last.

But a title would be needed. In 1950 my wife's mother, Florence Ada Keynes, in her ninetieth year, published a beautiful little book called *Gathering Up the Threads*. The title described very well what I had to do, and I might have been tempted to use it myself had I thought of it first. That being out, I had to look for something less romantic. The force–velocity relation had been in my mind for many years, at first in primitive form. It was linked later, as can be seen from the record, with the gradually evolving ideas of the series elastic component and the active state. They could be put together, with various intermediate things, under the title, *First and Last Experiments in Muscle Mechanics*.

There had, in fact, been many other experiments of mine before 1923 on a variety of themes; but none before 1920 of any serious import in relation to muscle mechanics. In 1920 I began to get interested, particularly for human muscles, in the relation between speed of movement and work performed. The results of that, never properly thought out, were revolving in my head when Gasser arrived. Our experiments in 1923–4 on *The Dynamics of Muscular Contraction* were indeed the start, as mine in 1964–7 were the finish, of the adventure described in this book.

A. V. HILL

11a Chaucer Road
Cambridge
December 1969

UNITS AND SYMBOLS

1. *The S.I. System.* In a recent publication the Royal Society urged that the system of units known as S.I. (*Système Internationale d'Unités*) should be adopted in all scientific and technical journals. Though this book is not a journal, I should have liked to do the same here; but was deterred when I contemplated the perplexities of readers who found muscles weighed in kilograms, having lengths in metres, developing force in newtons, doing work in joules, after stimulation by alternating current of so many hertz. Indeed, since muscle physiologists inevitably calibrate their instruments by hanging weights on them; since no property of living muscle can be measured more accurately than to one part in 1,000; and since the constant of gravitation varies little more than 0·1 % from place to place; for all such reasons it seemed to me that it would be a waste of time not to continue to measure muscle force in g wt., work in g-cm, heat either in g-cm or calories, and frequency in c/s.

In all the published papers referred to in this book quantities were given in such old-fashioned units. No advantage would be gained yet by using others, and some confusion might be caused.

2. *The micron.* It will probably be some years before physiologists and histologists, except under duress, cease to use their pet name micron, or μ, to mean 0·001 mm. Histologists began to use it 100 years ago, S. P. Langley (astronomer and spectroscopist) in 1883 for measuring wave-lengths, and Michael Foster in 1888, in his Text Book of Physiology (5th ed.), for the diameters of blood cells and nerve fibres. It is true that more recently μ has come to have a second meaning, namely 10^{-6}; but no more confusion need arise from this than from the double use of m for metre and 10^{-3}. Moreover, though it looks grand, there is no magic in expressing length in ångströms, and it is not everyone over 60 who can remember that these are really $10^{-4}\,\mu$.

3. *Quantities and symbols.* In dealing with the mechanics of muscle the generally accepted terms and symbols that follow are used without

further definition. The units are those employed unless otherwise stated. A few definitions are given at the end.

M	mass ('weight') of muscle, mg
P	force (tension) exerted, g wt.
P_0	maximum tension in isometric tetanus, g wt.
P/P_0	is often used for greater generality
v	velocity of shortening, mm/sec
v_0	maximum velocity under zero load
l	length of muscle in general, mm
l_0	'standard length', see below
t	time, msec, from a stated zero
t_0	time-constant of an exponential (in e^{-t/t_0})
x	length of contractile component of muscle, mm
y	length of series elastic component (s.e.c.), mm (the absolute values of x and y cannot be defined, only differences or differential coefficients are used)
a, b	the constants of the force–velocity equation,

$$v = b \ (P_0 + P)/(P + a)$$

S.D.	standard deviation of a single observation from a mean value
S.E.	standard error of the mean
l_0	for sartorii of frog or toad, is defined as the distance between the points where the muscle joins its pelvic and tibial tendons when the legs are stretched out in line perpendicular to the body. The sarcomere length when the muscle is at length l_0 is taken as $2 \cdot 25 \ \mu$
Pl_0/M and P_0l_0/M	are used to define the force exerted by a muscle in a more general form, account being taken of its dimensions. If P, or P_0, is in g wt., l_0 in cm, M in g, Pl_0/M is roughly in g wt./cm² of cross section. In a muscle in good condition P_0l_0/M is usually 2000 or more.

4. *Ringer's solution*, for frog and toad muscles, usually had the composition (mM): NaCl 115·5, KCl 2·5, CaCl$_2$ 1·8, generally with the addition of a phosphate mixture to give a concentration of 3 mM and a pH of about 7·0.

NOTE FOR THE READER

References are printed after each chapter, following the Harvard system, the date being given in the text after the author's name. If, in the text, no name is attached to a date mine can be inferred. No general list of references is given but names will be found in the general index.

Figure numbers indicate chapter as well; in ch. 5, for example, Fig. 5.1, Fig. 5.2,...

Chapters 3–9 are divided into sections, with titles; these are numbered §1, §2,... in each chapter. A list of chapters and sections is given under CONTENTS.

1

THE DYNAMICS OF MUSCULAR CONTRACTION

The title and most of the contents of this chapter are taken from a paper by Gasser and me in 1924. In choosing the title of the book the *first experiments* in mind were those we made together[1] at University College in the autumn and winter of 1923–4. The *last experiments* were made in 1964–7, also at University College; they have provided most of the experimental material for the rest of the book.

How it happened that Gasser came to work in London in 1923 is described in his autobiography (written characteristically in the third person) which reached Joseph Hinsey after his death in 1963. Believing that 'this autobiography should be made available to others', Hinsey had it published in *Experimental Neurology*, Suppt. 1, 1964. Like most good things that happen, Gasser's coming to University College seems almost accidental. Records which he had made in 1922, with the first cathode ray tube ever used for nerve physiology, had convinced him of the compound nature of the action potential, with velocity depending on fibre diameter. An International Congress of Physiologists was due to be held in Edinburgh in July 1923; to quote Gasser's words:

to satisfy a growing desire for an introduction to European physiology Gasser hit upon demonstration of records of the new findings as an excuse for attendance.

But in the mean time Abraham Flexner and Gasser's colleagues in St Louis intervened with a more extended plan lasting two years. This began, after the Congress, with six months in my laboratory, followed by periods with Walther Straub, Louis Lapicque and Henry

---

[1] I think Gasser really made them all, for I was particularly busy then; but we discussed them every morning for several months, and generally he had something new to report. He had never worked on the subject before and he never did again. But I doubt whether I should have made the experiments myself without him, certainly not so well.

Dale. Then only could he get back to the nerves and tube for which he had been homesick.

In his autobiography, referring to the 'long paper' which resulted from his experiments and our daily discussions, Gasser wrote:

When adherence to the schedule of a program brought about his departure both Hill and Gasser were in agreement that the paper contained much which would be called into question and much needing further clarification and revision.

True enough. What seemed to me at the time to explain our findings (and many that had preceded them), namely the visco-elastic theory of muscle contraction, was wholly wrong; and the experiments on the visco-elastic model are now irrelevant. But the theory took, like Charles II, an unconscionable time dying. Which is the more peculiar, because in my Nobel Lecture in Stockholm on 12 December 1923, I said:

It is clear that those of us who supposed that a stimulated muscle is simply a new elastic body were wrong, A muscle may possess some elastic properties but it requires more energy to do more work, a fact which is fundamentally in opposition to the elastic body theory.

That was spoken just before the first of Fenn's papers[1] was published, and six months before the second one. I knew Fenn's conclusions very well and they were obviously the death warrant of the visco-elastic theory; yet I went on thinking in terms of it for a long time after. It is odd how one's brain fails to work properly when pet theories are involved. Fortunately the theory did not affect the results of the experiments with Gasser, indeed it suggested some of them (see p. 15 below); and those experiments have been the basis of much that followed over 40 years—I hope with the 'clarification' that Gasser looked forward to.

In shortening this paper I have omitted references and much that no longer seems relevant; and have inserted a few comments to make things clearer or to preserve a consecutive argument. The comments are set in smaller type to distinguish them from the main text. The footnotes with my initials were added to compare our findings or ideas of 1924 with present knowledge.

[1] Fenn, W. O. (1923), *J. Physiol.* **58**, 175 and (1924) *ibid.* p. 373.

To the present-day reader the experiments described may seem primitive. They were made rather in a hurry because of Gasser's 'schedule', with simple equipment immediately available, no transducers, no amplifiers, no cathode-ray tubes, not even photography. But Gasser's simple little masterpiece (Fig. 3, p. 7), writing on a smoked drum, was able to produce records (Fig. 4, etc.) which can be recognized, after 45 years, as containing much of the information which later and easier methods have made the subject of current discussion.

THE DYNAMICS OF MUSCULAR CONTRACTION

By H. S. Gasser and A. V. Hill

Proc. Roy. Soc. B (1924), **96**, 398–437

Introduction

This investigation started with an attempt to verify, on isolated frog's muscle, the relations established recently in human muscles between speed of shortening and work done. When the flexor muscles of the arm are set to exert a maximal voluntary contraction, the velocity with which they shorten can be varied by altering the inertia of a load to which they are opposed; the work done is then found to decrease as the speed of shortening is increased. Since the extent of the movement in these experiments is always the same, if the work varies in any regular manner, so also must the force exerted by the muscle during any element of the shortening; *hence the force exerted by the muscle decreases as the speed of shortening increases.* This variation of work (or force) with speed is an important factor in determining the mechanical efficiency of muscle. Just as important, however, is its bearing upon theories of the nature of muscular contraction, and upon the problem of the adaptation of the muscles of different animals to the speed required of them by the dimensions and habits of their owners.

It was possible that the connexion between force exerted and speed of shortening might depend on some automatic regulatory mechanism inherent in the nervous system. The rapid shortening of a muscle opposed to a small inertial load might cause a proprioceptive reflex, withdrawing or reducing the volley of nervous impulses directed at the fibres of the muscle in consequence of what was intended to be a maximal voluntary effort. This could clearly be decided by repeating the experiments with an isolated muscle stimulated directly. Part I is an account of experiments which prove that the same general type of relation between work and speed of shortening occurs in isolated muscle. Hence the phenomena investigated

4

are not due primarily to an intervention of the nervous system, but to some fundamental character of the muscle fibre itself. Part II contains an account of experiments on the effect produced, by the speed of shortening or lengthening, on the force exerted by an active tetanized frog muscle. The speeds investigated varied from very rapid to very slow. The effects are quite different in magnitude in an active stimulated muscle and in a stretched inactive one. The active muscle behaves as though it possesses a much greater viscosity than an inactive one. This phenomenon is investigated in Part III, where the damping effect of a muscle upon a vibrating spring to which it is connected is discussed.

The effect of speed of movement on the force exerted by a muscle has been supposed to be due to the presence of an elastic network containing a viscous fluid; any change of shape of the muscle would require the viscous fluid to pass, more or less rapidly, into a new configuration relative to the solid structures of the tissues. In Part IV an elastic-viscous model is described, with which the phenomena discussed in Part II, for the case of the active muscle, may be imitated experimentally.

Hitherto we have dealt only with the properties of a muscle while fully 'active' during a maximal tetanic stimulus. In Part V are discussed certain changes in its elastic condition which occur in a muscle during the cycle of a single twitch. It is found that the rigidity and viscosity are at a maximum very shortly after the stimulus and then fall off continuously until contraction ends. The mechanical response, therefore, is fundamentally a very sudden phenomenon, its external manifestation—tension—depending on the flow of certain elements of the muscle into a new configuration necessary before that tension can be exerted.

Part I. The relation of the speed of shortening of an isolated muscle to its ability to perform external work

The apparatus in the rather elegant figure allowed one to vary the equivalent mass over a fairly wide range, and therefore the speed of shortening of the muscle. For a given distance of shortening the work was greater the less the speed—as it had been with human muscles. Everything else in Part I is omitted.

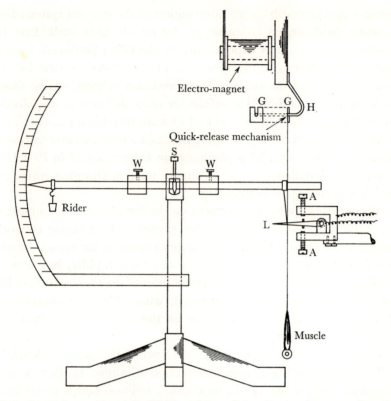

Fig. 1. Inertia lever with weights W, W, and screw S for adjusting the centre of gravity on the line of the knife edges. Rider and scale for measuring work. Lever L with adjustable stops, A, A, for limiting the extent of contraction and electrical timing of the duration of shortening. *Top*, quick-release mechanism. (The equivalent mass could be varied by adjusting the distance between the weights, and the point of attachment of the muscle to the lever.)

Part II. The effect of shortening or lengthening at various rates

(A) *The quick-release*

A muscle was stimulated with a tetanus; after reaching maximum tension it was allowed to shorten suddenly and its redevelopment of tension was recorded. This experiment was performed with a tension lever, the high frequency of which was obtained by choice of suitable dimension (see Fig. 3). A watch spring was soldered across the ends of a U-shaped brass bar. At its middle point was soldered a minute

6

projection P for the attachment of the lever. After mounting, the spring was ground down upon an emery wheel, until by trial its sensitivity was suited to the muscle studied. For recording, a *short*

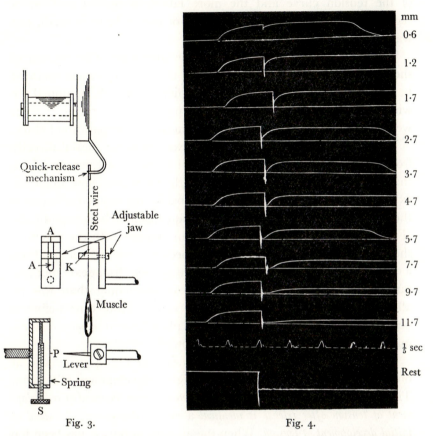

Fig. 3.

Fig. 4.

Fig. 3. Isometric tension recorder, with projection P for carrying short bamboo pointer and screw S for keeping spring taut; adjustable jaw for regulating amount of movement of knot K; A, A, slit guides for steel wire connecting muscle to quick release mechanism.

Fig. 4. Effect of quick release on tension recorded in maximal isometric tetanus. Frog sartorius. Right, amount of release, mm. Bottom, release of resting muscle under tension. (These records were made at room temperature, and so throughout the paper.)

piece of bamboo was used, about 3 cm long and just rigid enough not to vibrate of itself. The resulting lever had a usual undamped frequency of 350 per second, though one had a frequency of 450 per second. When a muscle is suddenly released during a tetanus the

7

vibration of the lever damps out in one to three vibrations. This damping out can be seen immediately after release in Fig. 4. Since the period is very short there is no serious interference with the form of the tension curve and the more laborious optical method of registration is avoided. The sensitivity of such a lever is small, but was found to be adequate.

The muscle was suspended by a fine steel wire. The amount of shortening was controlled by a knot K in the wire, which operated between two adjustable brass jaws in which the slots A, A, guided the wire but did not allow the knot to pass. Two sartorius muscles were used with the pelvis detached to make the preparation lighter.

When the tetanized muscle was released it never passed directly from the isometric tension corresponding to the greater length to that corresponding to the smaller length; no matter how small the extent of the release the tension fell below the isometric value of the shorter length and then rose again, approaching it asymptotically (Fig. 4). The fall in tension increases in amount with the extent of shortening allowed, until the latter becomes 10–15% of the total length, at which point the whole of the tension disappears momentarily on release.[1] If larger amounts of shortening than this be allowed, then the tension remains at zero a finite time before it begins to rise again, the duration of the phase of zero tension increasing with the amount of shortening allowed.

The curves of redevelopment of tension in Fig. 4 strike one at once as being similar to those of the original development of tension. This similarity suggests that the *process determining the speed of redevelopment of tension after release is identical with that determining the speed of initial development of tension.* When the amount of release has been too great, the tension not only falls to zero, but slack has to be taken up before tension redevelops.

When a *resting* muscle is released between two lengths at both of which it is under tension, it returns almost immediately to the tension of the shorter length (Fig. 4 bottom record). The maximum initial fall below the tension at the shorter length is slight, and the equilibrium tension is rapidly approached. This is true for releases made

[1] The irreducible amount, set by the series elastic component of the muscle, is now known to be about 3%; the compliance of the connexions and lever made up the rest. [A.V.H.]

over the whole range in which the resting muscle shows tension. The phenomenon described is due to some factor highly developed in the stimulated muscle, only slightly developed in the resting one.

(B) *The effect of speed of shortening upon tension*

These experiments were performed in two ways. In the first method the rate of shortening was controlled by a reducing gear made for controlling the speed of a motor-driven kymograph. It was provided with a clutch which ensured a start at maximum velocity. The

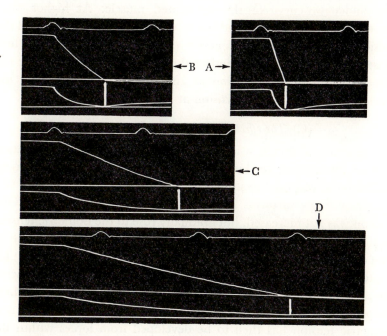

Fig. 6. Effect on force exerted by an excited muscle of allowing it to shorten at various speeds. Top line, time ($\frac{1}{5}$ sec); middle curve and base line, shortening; bottom curve and base line, tension. A most rapid shortening, D slowest, B and C intermediate.

motion was translated to the muscle through fine wire connexions and a bell-crank lever, the amount of movement being controlled by the brass jaws (Fig. 3). The rate of shortening was practically constant. In the second method the speed of shortening of the muscle was controlled by the arm of a lever whose movement was determined by a strong rubber band working against a resistance which was made by attaching the lever to the piston of an Albrecht piston recorder. The resistance, and therefore the rate of movement, were readily controlled by a glass stop-cock at the inlet.

Some records made by the second method are shown in Fig. 6. The upper line was traced by the controlling lever and shows the rate of shortening; the lower is the tension curve written by the lever to which the muscle was fixed at its lower end. These experiments show that, when the rate of shortening is sufficiently low, the tension at every length is equal to that which could be read off from the tension–length curve of the stimulated muscle. At higher rates the tension is always less, and before the shortening is completed the tension falls below the full isometric tension of the shorter length and must return to it after the end of the contraction. The greater the speed, the smaller the tension at any length, and the sooner the tension falls below the full tension at the short length.[1]

(C) The effect of forcibly extending a muscle

The amount of stretch was controlled by a knot in the suspension wire of the muscle, working between the two brass jaws. A strong rubber band was used to stretch the muscle. The sudden stretch of a tetanized frog's sartorius showed the surprising results[2] that the tension was not increased, but gradually rose to the isometric tension of the greater length, over a curve whose shape was the same as that of the initial development of tension (Fig. 8, 2., 5., 6.). The quick stretch causes a wave of tension to pass over the muscle, producing the first fling of the lever, but this tension is not maintained. It drops back to the isometric tension at the shorter length. The new isometric tension has then to be developed from forces within the muscle.

When the muscle is slowly stretched the result is quite different. If the stretch be sufficiently slow the tension follows the equilibrium relation of the tension–length diagram. For more rapid stretches

---

[1] Here surely was an invitation to look for an explicit force–velocity relation; but nobody looked for one till Fenn and Marsh in 1934. [A.V.H.]

[2] This is not so surprising when viewed from the standpoint of the sliding filament theory. In records 2., 5. and 6. the muscle started very short (estimated sarcomere length $1 \cdot 4$–$1 \cdot 5 \mu$) with its tension low, so the thin filaments were piling up at the Z lines. Anything might happen, one would think, when the thick and thin filaments were suddenly pulled apart. The records show what actually happened. [A.V.H.]

the tension at every point is *greater* than the equilibrium tension for the length, and when the stretching movement is arrested the tension falls back to the equilibrium state, rapidly at first and then approach-

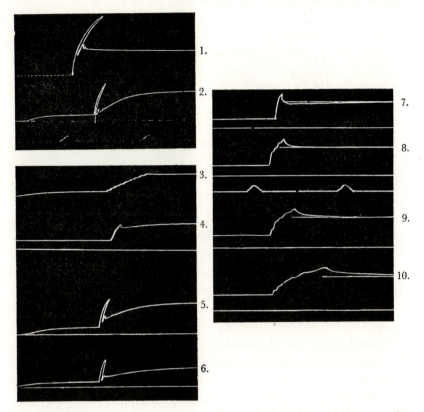

Fig. 8. Effect on tension in maximal tetanus of a stretch of varying quickness: 1., resting muscle for comparison, quick stretch; 2., 5. and 6., active muscle quick stretch; 3., very slow ('reversible') stretch; 4., fairly quick stretch; 7., 8., 9. and 10., in order, fairly quick to fairly slow stretch. Frog sartorius; different muscles were used for 1. and 2., for 3. to 6. and for 7. to 10.

ing equilibrium asymptotically (Fig. 8, 8., 9., 10.). One thus obtains an effect similar to that observed in the shortening curves but of the reverse sign.

When, however, the rate of stretch is still further increased another phenomenon, the one described above, appears. The results of such an experiment are recorded in Fig. 8, 4., 5., 6., 7. The tension gained by the stretch is not maintained, but falls off and must then be redeveloped by some internal process. The amount of this fall and redevelopment increases with the speed of stretching. The changes may be shown schematically in the diagram of Fig. 9.

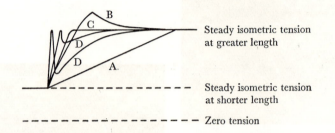

Steady isometric tension at greater length

Steady isometric tension at shorter length

Zero tension

Fig. 9. Effect of stretch at various speeds during maximal tetanus. A, very slow ('reversible'); D, rapid (upper), and very rapid (lower); B, fairly rapid; C, intermediate between B and D (upper). Frog sartorius.

Four types may be defined. In that characterized by A the tension is equal throughout to the equilibrium one for the length at the moment; in B the tension rises above equilibrium and returns to it asymptotically; in C the tension just attains the maximum value; in D, D some of the tension gained by stretching is lost and must be restored by forces within the muscle.

A muscle does not withstand sudden stretches well.[1] After two or three stretches the isometric tension developed at any length is often largely decreased, and in each successive tetanus there is a further decrease in tension, indicating that the muscle is being injured by the procedure.

When the unstimulated muscle is suddenly stretched (Fig. 8, 1.) the higher tension is promptly attained. The initial rise above equilibrium is slight and the return to equilibrium prompt. The effect is similar to that produced by the quick-release, but with the reverse sign.

[1] I suspect that this applies only, or particularly, to stretches (as described) from a very short length. I have made many stretches from length about l_0 without harmful results. Perhaps an excessive stretch may have helped to cause it. [A.V.H.]

(D) *Quick release followed by a quick stretch*

When a tetanized muscle rapidly shortens between two fixed lengths its tension drops to zero, as described above, although the equilibrium tension is more than zero at the shorter length. The question arises as to whether there is any change in the properties of the muscle thereby produced. Has the muscle lost its tension merely mechanically? Or has its sudden shortening somehow abolished its physiological activity? This could be tested by allowing the muscle to shorten, then quickly drawing it out again. The devices for stretching and releasing were combined therefore under the control of two knock-down keys, so that the two changes could be separated by any desired interval. The stretch was retarded by the piston-recorder, so that the tension attained on stretching would be maintained. When the muscle was released, then immediately stretched out again, the tension was promptly restored (Fig. 10, 2.), showing the muscle to be just as capable of stress as before its release; the physiological activity due to stimulation had not been abolished.[1]

(E) *Quick stretch followed by quick release*

A tetanized frog sartorius was stretched from a length of 23 mm to one of 27·5 mm, the extension being slightly retarded by the resistance of the piston recorder, so that the tension would rise above its equilibrium value. If the muscle be released as soon as the maximum tension is reached, the tension falls but little below the initial tension (Fig. 10, 5.). If the interval between stretch and release be longer the tension falls farther below the equilibrium tension on release (Fig. 10, 6.), and the fall increases with the interval until it is the same as that occurring in the quick release not preceded by a stretch (Fig. 10, 1.). The equilibrium tension is then approached from below. This suggests that when the muscle is stretched, if it be promptly released the antecedent state is promptly restored; but that if it be held under strain (Fig. 10, 3.) a change gradually occurs in it which makes it behave more like the muscle in the equilibrium state of the new length.

[1] A similar experiment was made by R. C. Woledge in 1961; *J. Physiol.* **155,** 187; Fig. 10B. [A.V.H.]

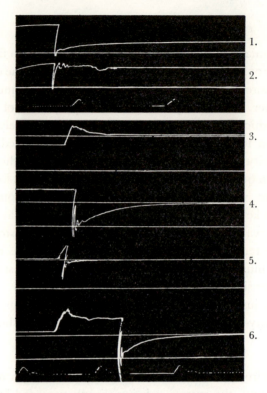

Fig. 10. Effect of quick release followed by quick stretch, and vice versa: 1., quick release only; 2., quick release followed by quick stretch; 3., quick stretch only; 4., quick release only; 5., quick stretch followed by quick release after shorter interval; 6., quick stretch followed by quick release after longer interval. Bottom, $\frac{1}{5}$ sec.

Part III. The viscosity and elasticity of stimulated muscle

This contains the beautiful but misleading experiment of Fig. 11; the rest of Part III is omitted. The effect shown cannot really be ascribed to a change of viscosity between rest and activity.

Part IV. An elastic viscous model

Everything that is now known makes nonsense of the visco-elastic theory and model; they had better be forgotten. Though they did lead us to make the experiments on single twitches illustrated in Figs. 17 and 18 below, on which the generalization of Fig. 16 is based. With minor modifications the diagram of Fig. 16 represents the current view today; see ch. 5, §1.

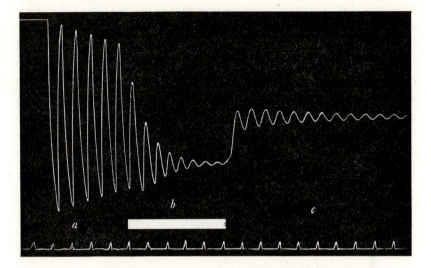

Fig. 11. Damping of oscillations in a spring connected to a muscle; *a.*, unexcited; *b.*, excited, *c.*, unexcited again. The damping becomes enormously greater when the muscle is excited. Time marks, $\frac{1}{5}$ sec.

Part V. The fundamental nature of the mechanical response to a single shock[1]

If the speed of development of the external response be determined by viscosity, it is desirable to know something of the time relations and characteristics of the internal mechanical change lying behind it. In a muscle twitch the 'initial' heat production occurs (*a*) in the early stages of the mechanical response and (*b*) fairly late in relaxation. The fundamental change, therefore, causing the external response in an isometric twitch is presumably completed quite early; the heat production associated with relaxation is probably nothing but the potential energy of the excited muscle degraded into heat. It would seem natural, therefore, to suppose that the fundamental mechanical change associated with contraction follows a course such as that sketched in the full line of Fig. 16, a sudden rise followed by a gradual fall. The dotted line shows the external mechanical response, which lags behind the internal change.

The chief means at our disposal for investigating the internal

[1] I have left in some of the viscosity jargon in order to show how we were led to the experiments of Figs. 17, 18 and 19, and to the ideas behind Fig. 16. [A.V.H.]

15

mechanical condition of the muscle is that of quick release or sudden stretch. If we apply a sudden stretch to a muscle at the moment A (Fig. 16), we should expect to be able to save it from the necessity of shortening internally under its own activity, before it can record an external tension; in terms of our model we should stretch out the network, which otherwise would have to shorten of itself in a viscous medium before becoming taut. Consequently we might expect to find a large rise of tension on stretching the muscle; we should be stretching a body which is less extensible and more viscous. A similar result might be expected at B. At C, on the other hand, where

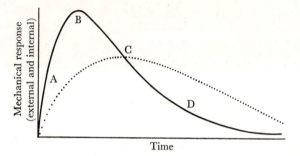

Fig. 16. Diagram of supposed internal ('fundamental') mechanical change in muscle during a twitch. Dotted line, observed external response lagging behind the internal change. A, B, C, D, see text.

the external tension recorded by the muscle is a maximum, a smaller rise of tension should occur as the result of stretching; the internal rigidity and viscosity of the active muscle have partly disappeared, and there is less left, so to speak, to stretch. At D, in the middle of relaxation, nearly all the extra internal viscosity and elasticity characteristic of activity have disappeared, and the remaining tension recorded externally by the muscle is mainly due to the fact that its internal structure has not yet flowed back to its initial condition. A sudden stretch at D, therefore, should produce little rise of tension— it should merely assist the muscle in its return, against viscous forces, to its resting state.

These expectations are strikingly borne out by the following experiments. A shock is applied and at any desired moment afterwards the muscle is subjected to a moderately rapid stretch, occupying about 0·01 sec. The result is shown diagrammatically in Fig. 17, while actual records are given in Fig. 18. In Fig. 17 a stretch at an early moment such as A leads to a large rise of tension. On the other hand, at C, where we might have expected the muscle (whose tension is now a maximum) to be at its most inextensible stage, the quick stretch produces much less effect. The extensibility has largely returned. At D the quick stretch produces little effect—the extensibility is rapidly reaching its resting value again.[1] In this way, by

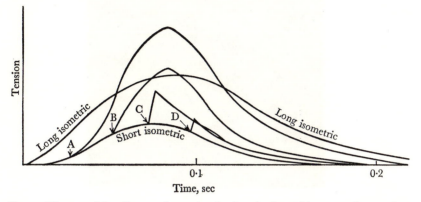

Fig. 17. Diagram of the effect, on the tension developed, of stretching a muscle at various moments during a twitch: copied from records.

quick stretches applied at various moments in a twitch, it is possible to track down the changes in elasticity resulting from stimulation before they have had time to affect the tension manifested externally. In a general way the result is that shown in Fig. 16. The stage of diminished extensibility rises more rapidly and falls more rapidly than does the tension manifested in an isometric lever attached to one end of the muscle.

The time relations of the physical change in the muscle can be seen in Fig. 18. When the stretch, which required 9 msec for completion, was made before or during the stimulus, the tension curve was unaffected. When the stretch started after the stimulus, an increase of tension occurred, the effect being greatest when the stretch started quite early after the stimulus. Clearly the muscle becomes very inextensible just after the latent period.

[1] See ch. 9, Figs. 9.1 and 9.2, for experiments made forty years later with very different equipment; and with much shorter stretches. [A.V.H.]

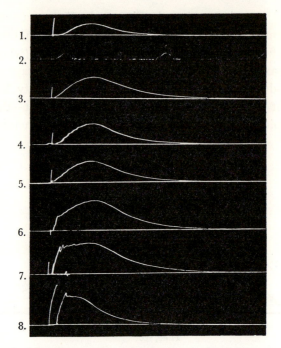

Fig. 18. Records of the effect, on the tension developed, of stretching a muscle at various moments during a twitch: 1., isometric (short); 2., ⅕ sec; 3., isometric (long), then stretches; 4., just before stimulus; 5., at stimulus; 6., 5 msec after stimulus; 7., 9 msec after; 8., 15 msec after.

Quick release during a twitch

When the converse experiment is performed, that is, when the muscle is released at various moments during a twitch, the findings support the conclusions drawn from the stretch experiments. A release made before or during the stimulus (Fig. 19, 6.) was without effect, while a release at the beginning of a twitch (Fig. 19, 2., 8.) caused the tension to fall and the whole twitch to progress at a tension below the isometric curve for the shorter length. In releases made when the twitch is near the crest the fall is greater (Fig. 19, 3., 11.).

After a release in the early portion of the twitch, the tension rises again after its fall, and the crest, when it does appear, is delayed. This redevelopment of tension is possible for releases occurring as late as the crest, in which case the second crest is much later than the first one (Fig. 19, 11.). Releases during the period of falling

18

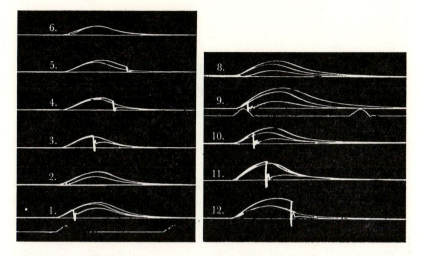

Fig. 19. Records of the effect, on the tension, of releasing a muscle at various moments during a twitch. Upper continuous line, isometric (long); lower continuous line, isometric (short); moment of release from long to short shown by break in curve. Order, from early to late release: 6., 2., 1., 3., 4., 5.; 8., 9., 10., 11., 12.

tension do not show a secondary rise, the maximum tension occurs just after the lever oscillation, and then the tension progressively falls as it does in the normal curve. In the lowest tracing in Fig. 18 of a 'stretched' twitch the crest is much earlier than the normal crest; in the released twitches it is delayed to various degrees.

The six pages that followed have no present significance and are omitted.

2

QUAMQUAM RIDENTEM DICERE VERUM QUID VETAT?

(Horace's Satires, I, i, 24)

Which could mean 'yet, what's to stop you from describing your experiments with a laugh?' The laugh is in curves A., B., C. and D. of Fig. 2.1; which do not look at all like four consecutive isometric contractions of a frog sartorius at o °C, all recorded at the same muscle length. They are used here to introduce some of the amusing problems discussed in later chapters.

The records were all made during maximal tetanic stimulation at length $l_0 - 0.4$ mm (l_0 was the 'standard length' of the muscle, 29 mm). Tension is given vertically as P/P_0, P_0 being the maximum tension developed; time horizontally in milliseconds from the start of each record. The curves are not shown above $P/P_0 = 0.6$.

About 0.5 sec before each record (A.–D.) began, stimulation started, at lengths respectively 1, 3, 4 and 5 mm greater than the final length l_0–0.4 mm. Contraction, initially isometric, was followed by rapid release to $l_0 - 0.4$ mm. The velocity of release, $4.1 \, l_0$/sec, was about twice as great as the intrinsic speed of the fastest fibres of the muscle. Release was ended by a stop, and the origin of time in Fig. 2.1 is the moment of 'stop' in each case. After that, while stimulation continued, the tension redeveloped isometrically.

In A. (1 mm release) the remaining tension at the moment of 'stop' was $0.068 \, P_0$, and redevelopment started at once along a curve concave downwards all the way. For B. (3 mm release) there was an interval of 9 msec before tension began to reappear; in C. (4 mm) the interval was about 19 msec; in D. (5 mm) the interval was about 24 msec.

The circles, every 5 msec, are for an ordinary isometric contraction, starting from previous rest; stimulation began at the origin and the latent period was about 15 msec.

The likeness of curves B. and C. to the ordinary isometric curve is striking—but misleading. In the isometric record the curvature at first is concave upwards, which is due to the fact that the 'active state' was not fully developed until about 50 msec after the start of stimulation. In the four curves A.–D. the active state had reached its plateau more than 0·4 sec earlier. The release was very rapid; during its course none of the muscle fibres which later produced records B.,

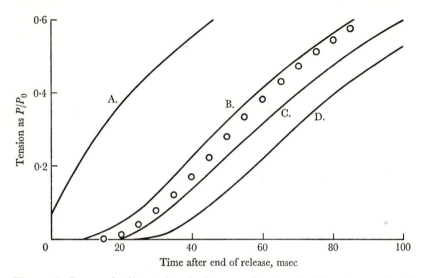

Fig. 2.1. A.–D., records of isometric redevelopment of tension of a fully active muscle; all at the same length ($l_0 - 0·4$ mm) after rapid release from maximum tension developed at various greater lengths. See text. The circles, every 5 msec, are for an ordinary isometric contraction starting from previous rest; stimulation began at $t = 0$.

C. and D. was able to keep up. The fastest fibres gave their first sign of redeveloping tension immediately after they had taken up their slack. The slower ones, which had been left further behind, were longer in taking up theirs, and only gradually joined the faster ones in developing tension.

The maximum slopes, dP/dt, of these curves are interesting. In units of P_0/sec they are:

A.	B.	C.	D.	isometric
18·9	10·1	9·0	9·0	10·9

Why is the maximum slope in A. so much greater than in the rest? The difference has been amply confirmed. In 16 experiments, during immediate redevelopment (like A.) the maximum value of dP/dt averaged 17·2 P_0/sec (S.D. of a single value $\pm$3·1 P_0/sec). In 16 ordinary isometric contractions the maximum value of dP/dt averaged 9·8 P_0/sec (S.D. $\pm$1·8 P_0/sec). In many delayed isometric contractions (ch. 4, §2) the slopes were usually 9 to 10 P_0/sec. The answer, believe it or not, is that the series-elastic component is stiffened during the development and maintenance of maximum tension, and this extra stiffness does not disappear at once during rapid release; but it does disappear when the tension is allowed to remain zero even for a very short time. There are more things in heaven and earth, Horatio, . . . and even in frog muscles.

3

THE FORCE–VELOCITY RELATION IN SHORTENING MUSCLE

§1. Introduction

Laulanié (1905) was among the first to recognize explicitly that in human muscular movement the efficiency (work/total energy used) varies with the speed. This was confirmed by Benedict & Cathcart (1913) and by many others since.[1] The existence of such a relation clearly invited one to look for another one between shortening speed and force exerted. Gasser and I (ch. 1, Part II B) were getting rather close to this in 1923 but we never got any closer; and a clear statement of it had to wait twelve years till Fenn & Marsh published their paper (1935). Apparently we were all hypnotized by the obvious importance, in man, of relating energy used to mechanical work performed at varying speed. It would have been easy at any time (at least with isolated muscles) to determine the elementary relation between force and velocity: which would have made things look much simpler.

Three years later than Fenn and Marsh, while working on the heat production of isolated frog muscle, I stumbled on another form of the force–velocity relation (1938). Its curve had empirically almost the same form as Fenn's but obeyed the simpler and more convenient equation:

$$v = b \ (P_0 - P)/(P + a).$$

Here P is force, v is velocity of shortening and P_0 is maximum force at zero speed (all these being measured); while a and b are constants chosen to give the best fit of the equation to a series of observed values of v and P. The constant b has the dimensions of velocity, a the dimensions of force. The chief interest of these, at first and for many years, was that a appeared to be the heat of shortening per centimetre. So $(P + a)v$ would be the rate at which the muscle was liberating

---

[1] References to some of this work, from 1922 onwards, are in Section C of my book (1965): C5, C7, C11, C22, C31, C34.

total energy during shortening, as work and heat together. It was, in fact, while investigating the total energy rate that I came on the equation. It appeared to be proportional to $(P_0 - P)$.

The maximum velocity of shortening, under zero load, must, if the equation holds over the whole range, be $v_0 = bP_0/a$. A usual value of a/P_0 in a frog sartorius is 0·25, so v_0 would be $4b$ (but see §4, ch. 4).

Actually in 1964 I found in frog muscle that the heat of shortening is not constant but depends on the load, being equal to a when $P/P_0 = 0·5$, greater than a when $P/P_0 < 0·5$. This may not happen in muscles other than those of the frog, though that seems rather unlikely. Certainly, however, with tortoise muscle Woledge (1968) found no evidence that the heat of shortening varies with the load; but the constant a of the equation was not equal to the heat of shortening per centimetre.

In the equation a has the dimension of a force, so a/P_0 is a number without dimensions; it would be expected therefore to be independent of temperature, which (nearly at least) is true. What is striking about it, however, is that it has about the same value over a wide range of animals and muscles. The b constant has the dimensions of a velocity and so could be expected to be largely increased by a rise of temperature; which again is true. For comparison between different muscles and animals b is best expressed in terms of b/l_0, where l_0 is the standard length of the muscle (or its fibres). Then its dimensions are $(time)^{-1}$ and it is found to vary enormously according to the size and habits of the animal and the functional rôle of the muscle.

The force–velocity relation and its equation have now come into current use. For example, Close (1964) gave a number of excellent force–velocity curves for fast and slow muscles of the rat recorded *in vivo*. His values of a/P_0 were regularly around 0·25; as were those of Ritchie (1954) with rat diaphragm. Close & Hoh (1967) reported similar results with the muscles of new-born kittens. Toad muscles have been widely used and there is a good force–velocity curve for one in Fig. 8 of my paper (1949). There is also much incidental information in that paper about the heat of shortening in twitches of toad muscles. Wilkie (1950) obtained good curves with human muscles,

In only one animal so far investigated has a value of a/P_0 been found greatly different from 0·25, far outside the range of possible variation. The exception is the muscle of the tortoise, *Testudo graeca* or *mauretanica*. According to Katz (1939) its force–velocity curve is of the usual type, but more curved, corresponding to a mean value of $a/P_0 = 0·11$ in eight experiments (temperatures between 0 °C and 20 °C). The low value of a/P_0 in tortoise muscles was confirmed by Woledge (1968), who in eight experiments at 0 °C found a mean value of 0·072 (S.E. of mean $\pm$ 0·008); his mean value of b/l_0 was 0·016 (S.E. $\pm$ 0·0026) which is about one twentieth of the mean value for frog muscles (see § 3). Woledge's interesting discussion of the energetics of tortoise muscle was pivoted on the unusual shape of the force–velocity relation, which is associated with the low value of a/P_0. This he found to be related to such important characteristics as the high ratio of mechanical work to (work + heat). With the more usual force–velocity relation a muscle is faster but less economical; in a tortoise, self-preservation depends less on speed than on armour, and its way of life makes economy in the use of energy vital.

The force–velocity relation has recently been subjected to a theoretical scrutiny by Caplan (1966), based on the methods of irreversible thermodynamics (see also Wilkie & Woledge, 1967; Woledge, 1968). On certain assumptions about the types of energy converters involved, a force–velocity relation of the usual kind could be deduced with only minimal reference to known properties of muscle. This would be wonderful, if confirmed, but Wilkie and Woledge have pointed out some serious snags in the assumptions. No doubt discussion will continue. Some of the regularities about the force–velocity relation have seemed sometimes too good to be true, one has been proved to be; others have appeared too good not to be true. It is early, as yet, to guess in which category the applications of irreversible thermodynamics will finally appear. It may well be that the attempted application to muscle may supply an excellent test of the principles themselves of irreversible thermodynamics; the properties of muscle are becoming rather well known and it is the most universal of prime movers.

In addition to the equation discussed above, and Fenn's, two others have been proposed for the force–velocity relation, by Polissar

afterloads P the velocity v is found from the slopes of the recorded isotonic contractions; finally the maximum tension P_0 is obtained in an isometric contraction at the same (the initial) length. With obvious precautions to avoid the effects of friction and inertia this method was used by nearly everyone who has investigated the relation, since Fenn & Marsh down to the present day.

At *shorter lengths*, outside the range referred to, the maximum force is less; and Abbott & Wilkie (1953) concluded that the usual equation can still be applied provided that P_0 is replaced by the maximum force that can be exerted at the length at which v is measured. At *greater lengths* the parallel elastic component introduces a complication, it provides a diminishing contribution to P while the muscle shortens; and again the maximum force developed is less than P_0. Together these effects would probably make the force–velocity relation too complicated to be useful.

In the traditional method of determining the P–v relation the force P is taken as the independent variable, and v is measured for various P's. But it is possible to reverse the procedure, to take v as the independent variable and measure P for various velocities of release. This has some advantages, particularly at the end of the curve with high velocities. It is described below in §4.

It would be interesting to have a force–velocity relation made with a single fibre. A sighting shot at this target was fired in 1966 (p. 182) by Gordon, Huxley & Julian. Possibly the job would be easier if velocity were taken as the independent variable, and force measured at different velocities. Then the fibre would not have to move anything except itself.

It is possible to determine the force–velocity relation from a single recorded isometric contraction, provided that the tension–extension curve of the series elastic component is known. Also, a pair of delayed isometric contractions, one with and the other without a known added compliance, can be used to give not only the force–velocity relation of the contractile component but also the tension–extension curve of the series elastic component. All this is discussed in §5.

§3. The constants a/P_0 and b/l_0 of the force–velocity equation of the frog sartorius at 0 °C

There are many values of these constants, for various muscles, scattered in the literature since 1938; but not enough of them were made under comparable conditions, to allow mean values and variability to be calculated. The only exception is for frog sartorii at 0 °C. For these, I gave (1938, Table III) eleven values each of a/P_0 and b/l_0. The average for a/P_0 was 0·262 and the S.D. of a single value was 0·033 or 12·6%; for b/l_0 the average was 0·343 and the S.D. of a single value was 0·038 or 11·1%. In 1939 (Table II) Katz gave the means of twelve values of each, 0·26 for a/P_0 and 0·34 for b/l_0. He did not report the individual measurements, but gave the range of variation which was about the same as in my experiments. When average values were allowable I have used $a/P_0 = 0·25$ and $b/l_0 = 0·325$.

§4. Measuring the force during a constant pre-set velocity

It is often amusing and sometimes profitable to recall how and when ideas came into one's head. The reason why I made the experiments in §4 is that when we decided early in 1967 to move to Cambridge it was natural to think of people I used to know there before 1920. Among them was Horace Darwin (1851–1928), then Chairman of the Cambridge Scientific Instrument Company, who lived at The Orchard in Huntingdon Road. That house no longer exists; in its place is a new women's College, to which a granddaughter of ours was going that autumn. I remembered well many visits there to Horace Darwin, in which we discussed the design of instruments, general or specific; and I keep a vivid mental picture of him sitting in a deep arm-chair with a block of squared paper on his knees. He talked about his principles of instrument design, one of which was to turn them round, or inside out, or upside down, in fact to change them in any way that might conceivably make them work better. This principle is just as good for designing experiments; and in thinking of him it occurred to me that it might be useful, in experiments on the force–velocity relation, to reverse the independent and dependent variables; in fact to measure the force when a muscle was constrained to shorten at a constant speed, instead of the other way round.

The instruments were ready to hand for these experiments, but little time was left to make them; and they had to compete for priority with other experiments described in ch. 4, §2. Both sets could have been made much better if more time had been available; but they had to be made in three months or not at all. The ones on the force–velocity relation are worth putting on record because they show clearly what the original method never revealed, the sharp rise at the end of the curve in the region of very low tensions (Fig. 3.2 below). This is due to the presence of a limited number of fibres of high intrinsic speed.

The experiments were planned as follows. An isometric contraction was calculated with average constants, $a/P_0 = 0.25$, $b = 10$ mm/sec. In Table 1 the tension (as P/P_0) from which release was

Table 1. *Calculation made in preparation for the experiments described in the text and illustrated in Figs. 3.1 and 3.2*

Force (P/P_0)	0.02	0.075	0.15	0.25	0.37	0.5	0.65	0.8	0.9	1.0
Time (msec)	32	48	65	85	110	136	177	245	314	—
Velocity (mm/sec)	35	28	21	15	10	6.7	3.9	1.9	0.9	0

to start is given in the first row, the time to that tension (the moment of release) in the second row, the velocity expected at that tension in the third row. The muscle was to be stimulated at an initial length rather greater than l_0 and released about 3 mm at a suitable speed. The stimulus was started by the opening of a key, release was effected by a second key, a third key ended the stimulus; and the distance of release was set by a stop. The speed of release was adjusted on a Levin–Wyman ergometer to which the muscle was attached.

If these speeds were correctly estimated for an actual muscle, then the tension recorder carried by the instrument would give a set of horizontal lines. Otherwise the tension would rise or fall fairly quickly to a constant value. The actual velocity could be calculated from the interval on the record, start to stop, to check the setting. The constant velocity so obtained was plotted against the observed tension to give the force–velocity relation.

That in principle is what was done. The ergometer could be adjusted rather accurately to the desired speed. The sweep started when the ergometer was released, it ended when the moving arm made contact with the stop. The base line was recorded automatically when the muscle had relaxed at the final length. The full velocity was reached quite quickly. A few isometric records were made at intervals so that P_0 could be interpolated. The protocol could be altered during an experiment as actual results accumulated.

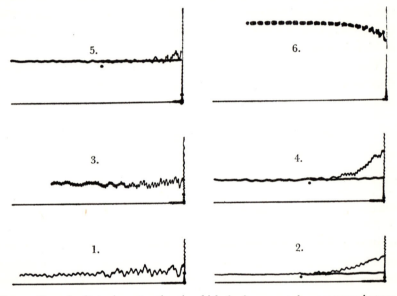

Fig. 3.1. Records of tetanic contractions in which the force exerted was measured at pre-set velocities. Tension vertically, time during release horizontally, one or two sweeps. Read from right to left. Frog sartorius, 0 °C. In records 2., 4. and 5. 'stop' is indicated by a dot. Details in the text.

In Fig. 3.1 are copies of six records from three experiments. They run from right to left. The vertical line at the right was due to the redevelopment of tension after release had ended. In record 1., with high gain (the vibration was due to this and the high speed), the resting tension at length $l_0 + 2$ mm is shown by the spot on the right. After the estimated time from the start of the stimulus, the ergometer was released at a moment which proved to be nearly right and little readjustment of tension occurred. After 57·5 msec the movement

(3 mm) ended at the stop, just before the end of a 60 msec sweep, and the beam was dimmed automatically and returned to the right, where it traced a vertical line. When the stimulus ended the tension fell and then after 2 or 3 sec the beam traced out the base-line. The tension (mm on the record) averaged 3·4 mm over the second half of the sweep; P_0 was measured as 151 mm, so P/P_0 was 0·022. The velocity of release was 52·1 mm/sec = 1·77 l_0/sec.

In record 2. with lower gain P/P_0 was 0·0567, the ergometer arm hit the stop in the middle of the second (60 msec) sweep (show by a dot), the velocity was 39 mm/sec = 1·27 l_0/sec. At the start the tension went rather too high (release had not been set quite early enough), but it had settled back to a constant level for the last 60 msec of the movement.

In record 3. with higher gain, P/P_0 was 0·049 and the velocity was 38·9 mm/sec = 1·32 l_0/sec. The start was almost exactly right (100 msec sweep).

In record 4., 3·5 mm release to $l_0 - 2$ mm started rather too late but the tension was constant for 100 msec before the end (at the spot). P/P_0 was 0·16 and the velocity 24·3 mm/sec = 0·79 l_0/sec (100 msec sweep).

In record 5. the tension was constant at $P/P_0 = 0·35$ for 160 msec before stop occurred in the middle of the second sweep (dot); and the velocity was 12·9 mm/sec = 0·46 l_0/sec (160 msec sweep).

In record 6. the muscle was released too early and the tension took some time to become constant; but it remained constant for about 250 msec before stop occurred. P/P_0 was 0·55 and the velocity was 7·4 mm/sec = 0·251 l_0/sec (500 msec sweep).

Twenty-four measurements made in these three experiments are shown together in Fig. 3.2. The quantities plotted are v/l_0 and P/P_0. The top point on the vertical axis is the mean velocity of shortening under strictly zero load (taken from ch. 4, §1). The curve is drawn from the usual equation $v = b \, (P_0 - P)/(P + a)$. No such equation could fit the observed points below $P/P_0 = 0·05$; above that, the best fit with an equation of this type was with constants $a/P_0 = 0·26$ and $b/l_0 = 0·42$. These give the nominal maximum velocity under zero load as $0·42/0·26 = 1·615$ l_0/sec, which is some way below the highest velocities observed. This, as is shown in ch. 4, is due to a significant fraction of the fibres being faster than the average.

The value of a/P_0 (0·26) used in calculating the curve is close to the average of a large number of values in the literature (see §3 above). But the value of b/l_0 (0·42) is rather unusually high, though

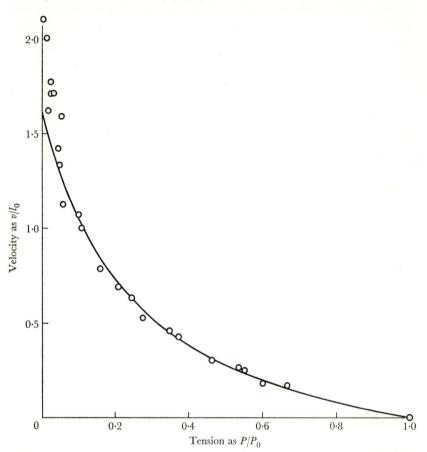

Fig. 3.2 Force–velocity relation, derived from experiments made with velocity as the independent variable. Velocity as v/l_0, force as P/P_0. Twenty-four measurements in three experiments are shown as circles. The curve was drawn from the usual equation with constants given in the text.

not outside the range of values given by me and by Katz (§3). One cannot argue from the mean for three muscles. Nevertheless, it is possible that velocities of shortening may be rather greater when obtained by this method than by the one used hitherto in which the muscle had to move a loaded lever; in the present experiments it had nothing to move but itself, together with the Ringer's solution dragged along as it shortened. But that is a matter for later inquiry.

In a frog sartorius at o °C the active state is fully developed after about 50 msec (ch. 5, §2). This is much less than the time in these experiments, even with the most rapid release, at which the velocity of shortening was measured.

§5. The force–velocity relation of the contractile component during an isometric contraction

A. *Argument*

During an isometric contraction the contractile component shortens and stretches the elastic elements in series with it. The amounts of shortening and stretch are discussed in ch. 7, §2; they can be calculated 1. if the force–velocity relation is known and 2. if the active state of the muscle is fully developed. If conversely the tension–extension curve of the series elastic component is known, and if condition 2. is satisfied, the force–velocity relation can be calculated. The argument is as follows.

Let x be the length of the contractile component and y that of the series elastic component. It is not necessary (or indeed possible) to know the absolute values of x and y, we are concerned only with differences or differential coefficients. Provided that the muscle has reached its fully active state,

$$dx/dt = (dP/dt)/(dP/dx).$$

Now $-dx/dt = v$, the velocity of shortening, and since the contraction is isometric

$$x + y = \text{constant}.$$

Therefore

$$v = (dP/dt)/(dP/dy).$$

But dP/dt is the slope, which can be measured, of the $P-t$ curve of the isometric contraction; while dP/dy is the slope of the tension–extension curve. Thus if dP/dt and dP/dy are measured at any P, their quotient is the velocity v at that P. Repeating this at various P's one obtains the force–velocity relation.

The practical utility, however, of this method is not great, because the tension–extension relation of the series elastic component cannot be determined directly except by the method of quick release from maximum tension as described in chapters 6 and 7; and the relation so obtained cannot be applied accurately except to a special kind of isometric contraction, viz., that of tension redevelopment after release from a high tension to a low but not zero tension. But it has a theoretical value in showing that a normal force–velocity relation can be obtained during a contraction in which tension is varying rapidly with time. This is of interest because it has been questioned whether velocity varies normally with tension when tension is changing quickly. This is discussed in §6 below, in answer to the query: Is the force–velocity relation an 'instantaneous' property of muscle? The conclusion reached there is that it is. The force–velocity curve now to be derived confirms that conclusion.

B. *Experimental*

In Fig. 3.3 are eighteen points (hollow circles) giving velocity v obtained as described and plotted against P/P_0. The contraction was an isometric redevelopment of tension *after* release from maximum tension (like curve A., Fig. 1.1); the tension–extension curve was obtained *during* rapid release from maximum tension, as in ch. 6, §2. These were not the same contractions because recording during release had to be very rapid, for tension redevelopment it was slower; otherwise they were comparable. The eighteen points are close to a curve (not drawn) given by the usual equation with $a/P_0 = 0.28$, $b = 0.388$ $l_0/\text{sec} = 12$ mm/sec. This is a perfectly normal force-velocity curve, it might have been obtained by either of the methods described earlier in this chapter.

In Fig. 3.3 also are nineteen points (filled circles) giving shortening velocity calculated in the same way, but from an ordinary isometric contraction starting from previous rest. dP/dy was the slope of the same tension–extension curve, while dP/dt was the slope of the isometric tension curve. The numbers under six of the points are times (in msec) from the start of stimulation. The filled circles do not provide a recognizable force–velocity curve, even at times greater than 50 or 60 msec by when the active state should have reached its plateau (ch. 5, §2). The large divergence at low tensions and short times is due to the active state not being fully developed; but at higher tensions and longer times the divergence is due to the tension–

34

extension curve, made by rapid release from a high tension, not being suitable for use with an isometric contraction from previous rest. This is discussed in ch. 7, §§2, 3 and 4. If the method were applied to a delayed isometric contraction (see ch. 4, §2), during which the muscle was fully active, the first reason for divergence would not hold; but the second one would, because the tension–extension curve would be unsuitable for use with a muscle which had not been stretched by a previous high developed tension.

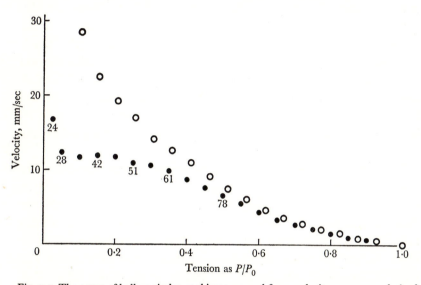

Fig. 3.3. The array of hollow circles, making a normal force–velocity curve, was derived from a single isometric contraction (redevelopment of tension after quick release) and a corresponding tension–extension curve. The array of filled circles was derived in the same way from an ordinary isometric contraction from previous rest. It does not provide a recognizable force–velocity curve. The numbers under some of the filled circles are times (msec) from the start of stimulation.

This is not because the tension–extension relation does not exist, but because it depends on the muscle's immediate previous history of stress–strain. It can be derived by the sort of analysis described in ch. 7, but only by the use of a force–velocity curve, so it would merely be arguing in a circle to find v from the ratio $(dP/dt)/(dP/dy)$; the force–velocity curve would already have been used in determining (dP/dy). The only method hitherto described of determining the tension–extension curve directly is that of rapid release from a high developed tension; but that makes the result unsuitable for any contraction other than isometric redevelopment after immediately prior release. Another method, however, is discussed in C below, which has the advantage of determining the force–velocity curve as well.

C. *Calculation from two delayed isometric contractions*

The suggestion was made in 1938 (referred to in ch. 7, §1 below) that two records of isometric contraction, one with and the other without a known added compliance, would allow the force–velocity curve to be calculated; it was not recognized originally that the tension–extension curve could be obtained too. Wilkie (1950) made some experiments in this way on human arm flexors, and Macpherson (1953) on frog sartorii. It was not realized by either that with small tensions (i.e. at short times) the active state of the muscles might not be fully developed; it would be more fully developed (at a given P) in the slower contraction, with added compliance, which would make the error from a comparison greater.

This uncertainty can be avoided by using delayed isometric contractions, as described in ch. 4, §2. For this purpose it would be simpler to use a preliminary isotonic contraction to provide the delay, and to stop it by the connexion to a tension recorder becoming taut. That has been tried and works. Records of tension would be made during the isometric phase and their slopes dP/dt, with and without an added compliance, measured at the same value of P. Let the added compliance be c (mm/g wt.) and let subscripts o and c be used to signify contractions without and with the added compliance. Then the equations are (cf. those in §5A):

$$v \, (dt/dP)_0 = (dy/dP)_0, \tag{1}$$

$$v \, (dt/dP)_c = c + (dy/dP)_c. \tag{2}$$

If it be assumed that $(dy/dP)_c$ and $(dy/dP)_0$ are the same:

$$v = c/\{(dt/dP)_c - (dt/dP)_0\}. \tag{3}$$

This gives v at tension P, so if the (dt/dP)'s are measured at a series of values of P, and v from eqn 3 is plotted in a curve against P, a force–velocity relation is obtained. If v, at any P, is substituted in eqn 1 one gets the value of dy/dP; this, against P, gives a convenient form of the tension–extension curve, but it can be integrated numerically if needed to give Δy as a function of P.

The only snag about the argument is that $(dy/dP)_0$ and $(dy/dP)_c$ may not be quite the same, in which case one must use eqn 4:

$$v = \{c + (dy/dP)_c - (dy/dP)_0\} \ / \ \{(dt/dP)_c - (dt/dP)_0\}. \tag{4}$$

It is believed that the series elastic component has a certain amount of hysteresis; under tension it goes on stretching (ch. 7, §5). The rise of tension in an isometric contraction with added compliance is slower than without it, so $(dy/dP)_c$ at any P would tend to be rather greater than $(dy/dP)_0$. If c is small enough the difference would be insignificant, but working with a small value of c would make the measurement of $(dt/dP)_c - (dt/dP)_0$ more difficult and less accurate. Though in any case it would be well to work with a value of c which is small enough to allow the final length of the muscle, at maximum tension, to be nearly the same with and without the added compliance.

A few preliminary experiments only have been made in this way, but their results were encouraging.

§6. Is the force–velocity relation an 'instantaneous' property of muscle?

The force–velocity relation of muscle has usually been determined by measuring the velocity of shortening against a number of constant loads; but, as described above (§4), the alternative method can be used of measuring the force with a number of constant velocities. In both methods constant quantities are set or measured, so neither can give any indication of whether the relation is an instantaneous property of muscle, or whether, when force and velocity vary with time, some interval is required before the contractile mechanism adjusts itself to a change. Jewell & Wilkie (1958, p. 524) concluded, from their observations, that in frog sartorii a change of velocity follows a change of force very quickly: 'probably in less than 1 msec, certainly in less than 6 msec'. Huxley's theory, however, appears to predict (1957, p. 304) a much greater time-lag.

Doubts have naturally been expressed as to whether the force–velocity relation is applicable to an isometric contraction, in which the rate of tension change is high and varies continuously throughout.

It was shown, however (§5B, Fig. 3.3), that an apparently normal force–velocity curve can be calculated from a record of isometric redevelopment of tension after release; though in a similar calculation, using a record of a normal isometric contraction starting from previous rest, serious differences were found which could be attributed: (a) to the fact that the active state takes time to develop and (b) to a temporary stiffening of the series elastic component under a high developed tension.

The clear result obtained with the curve of rising tension during redevelopment is the best evidence hitherto that when force and velocity are varying rapidly the relation between them is much the same as when they are not varying at all. But it does not prove that lag between change of velocity and change of tension is negligible. That cannot be great but it might be detected by other methods.

The difficulty about drawing conclusions from the external application of a rapid change, either of force or velocity, is that a muscle is mechanically buffered by its series elastic component, and this slows the resulting change of force. This is illustrated by the records (exact copies) in Fig. 3.4. For record A. (200 msec sweeps) a muscle was stimulated isometrically at length $l_0 + 0.5$ mm, and at 53 msec after the start (point a., $P = 0.215 P_0$) it was released at a speed of 24·5 mm/sec. The value of dP/dt just before release was 9 P_0/sec, immediately after release it was $-4.2 P_0$/sec; there is no visible lag in passing from one to the other. Then shortening continued and the tension fell nearly to the value corresponding to 24·5 mm/sec in the force–velocity relation. When release was suddenly ended after 1·5 mm by a stop (point b., $P = 0.126 P_0$), the value of dP/dt just before was $-0.18 P_0$/sec, immediately after it was $+15.4 P_0$/sec. Again there was no visible lag in passing from one to the other.

The base-line below was made automatically after the muscle had relaxed.

For record B. (160 msec sweep) a muscle was stimulated isometrically and released (point a.) at time 21·5 msec with a velocity of 33 mm/sec. The release (1·7 mm) was ended by a stop at 72·8 msec (point b.). Just before b. the tension had nearly reached a level ($P = 0.04 P_0$) corresponding, in the force–velocity relation, to 33 mm/sec. Immediately after point b. the tension started to rise rapidly. The slight apparent lag in reaching its maximum slope is attributed (ch. 4) to the high speed of release, which caused some of the slower fibres to lag behind; when they had taken up their slack, about 10 msec later, dP/dt reached its maximum and then gradually decreased.

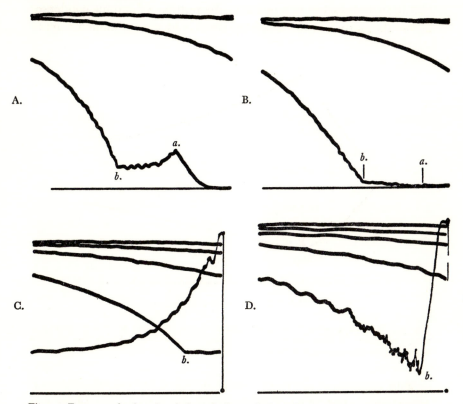

Fig. 3.4. Four records of tension, before, during and after release of a frog sartorius during stimulation: a., release; b., stop. Read from right to left. For details see text.

In record C. (100 msec sweep) a muscle was stimulated iso-metrically until its maximum tension was reached, then released 2·14 mm with velocity 18·8 mm/sec. Before the stop (at b.) the tension was still falling and dP/dt was small and negative; immediately after the stop the tension began to rise at its maximum rate, dP/dt was 12·4 P_0/sec. There is no trace of lag in the turn-over. The speed of release had been so low that none of the fibres had lagged behind.

In record D. (60 msec sweep) a muscle was released 0·77 mm at 120 mm/sec from previous maximum tension; release was ended by a stop (point $b.$, $P = 0·11\ P_0$). There is a lot of vibration in the record; this was due to the sudden end of a very rapid release. But it is not too difficult to draw a mean line through the maze of vibration and discern its true form. (By putting a damping condenser across the output of the amplifier one could have obtained a more

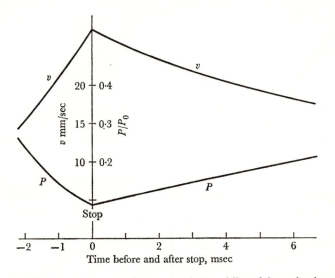

Fig. 3.5. Rapid release of tension during stimulation, followed by redevelopment: a magnified picture of what happens just before and just after 'stop'. Lower lines observed tension, upper lines calculated shortening velocity.

elegant record; but the more elegant it looked the less information it would give.) The value of dP/dt during the early falling phase was very large and negative; at the stop dP/dt changed sign to its maximum positive value, 16·3 P_0/sec.

Again there is no evidence of a discontinuity of any kind, except that dP/dt changes sign. A lag such as that which was just detectable at the start of redevelopment in record B. could not be seen through the vibrations, though with so high a velocity of release some must have been present. But the amount of release was so small that the slow fibres would soon catch up.

From another record of tension, falling during release and rising after stop, P was read off at intervals of 1 msec and is shown before and after stop in Fig. 3.5. dP/dt immediately before stop was -45 P_0/sec, after release it was $+19\cdot8$ P_0/sec. From P at any moment the velocity v of shortening of the contractile component was calculated from the usual equation with $b = 10$ mm/sec and $a/P_0 = 0\cdot25$. The time course of v is shown by the upper line. While P was falling rapidly before stop, v was rising rapidly; while P was rising more slowly after stop, v was falling more slowly. No discontinuity occurs at stop, either of P or of v, only of dP/dt and dv/dt. There is nothing novel about Fig. 3.5, but it may help the reader to visualize what is happening during some of the rapid changes described earlier.

References

Abbott, B. C. & Wilkie, D. R. (1953). *J. Physiol.* **120**, 214.

Benedict, F. G. & Cathcart, E. P. (1913). *Carnegie Inst. Pub.* No. 187, Washington.

Best, C. H. & Partridge, Ruth C. (1928). *Proc. Roy. Soc.* B, **103**, 218.

Caplan, S. R. (1966). *J. theoret. Biol.* **11**, 63.

Close, R. (1964). *J. Physiol.* **173**, 74.

Close, R. & Hoh, J. F. H. (1967). *J. Physiol.* **192**, 815.

Fenn, W. O. & Marsh, B. S. (1935). *J. Physiol.* **85**, 277.

Gordon, A. M., Huxley, A. F. & Julian, F. J. (1966). *J. Physiol.* **184**, 170.

Hill, A. V. (1938). *Proc. Roy. Soc.* B, **126**, 136.

Hill, A. V. (1949). *Proc. Roy. Soc.* B, **136**, 195.

Hill, A. V. (1964). *Proc. Roy. Soc.* B, **159**, 297.

Hill, A. V. (1965). *Trails and Trials in Physiology.* London: Edward Arnold.

Huxley, A. F. (1957). *Prog. Biophys.* **7**, 255.

Jewell, B. R. & Wilkie, D. R. (1958). *J. Physiol.* **143**, 515.

Katz, B. (1939). *J. Physiol.* **96**, 45.

Laulanié (1905). *Eléments de physiologie,* 2 ed. Paris; results quoted in Benedict & Cathcart (1913), p. 104.

Macpherson, L. (1953). *J. Physiol.* **122**, 172.

Ritchie, J. M. (1954). *J. Physiol.* **123**, 633.

Sonnenblick, E. H. (1962). *Fed. Proc.* **21**, No. 6, 975.

Sonnenblick, E. H. (1965). *Fed. Proc.* **24**, No. 6, 1396.

Wilkie, D. R. (1950). *J. Physiol.* **110**, 249–80.

Wilkie, D. R. & Woledge, R. C. (1967). *Proc. Roy. Soc.* B, **169**, 17.

Woledge, R. C. (1961). *J. Physiol.* **155**, 187.

Woledge, R. C. (1968). *J. Physiol.* **197**, 685.

4

THE STATISTICAL DISTRIBUTION
OF INTRINSIC SPEEDS IN THE FIBRE
POPULATION OF A MUSCLE

Definition. The intrinsic speed of a muscle, or fibre, is the maximum velocity with which it can shorten under strictly zero load. In a muscle in which a 'standard length' l_0 can be specified the intrinsic speed is expressed as a multiple of l_0/sec.

§1. The intrinsic speed under strictly zero load

It is not possible, by traditional methods, to measure the velocity of shortening under zero load. Without any applied tension an isolated muscle at rest shrinks, folds, or wrinkles, and its length becomes indeterminate. It was shown (1949, p. 427) that at lengths considerably less than l_0 a muscle at rest still exerts a finite tension. Moreover, any external device for recording movement introduces inertia and friction, however small, and tension has to be developed to overcome them. A fairly near approach, by normal methods, to shortening under zero load was described in 1951. Isotonic twitches of a frog sartorius under 1 g load (about 0·01 P_0) were recorded with rather high sensitivity and small inertia; the velocity of shortening reached 1·9 l_0/sec at about 24 msec after a shock applied 'all over'. The distance shortened by that time was a fraction of 1 mm.

No doubt the method could be improved, but it was really better to invert it as follows. During a tetanus a muscle was released from maximum to zero tension, at a speed greater than that of its fastest fibres under zero load. The release was continued 1 or 2 mm further and ended by a stop. On the record so obtained the time was measured to the moment at which tension just began to redevelop. This allowed the velocity of shortening of the fastest fibres to be calculated. The procedure has the advantage that the active state had been fully developed before shortening began; though in my experiments of 1951 the fast fibres that gave the high velocity were probably already fully active at 24 msec.

In Fig. 4.1, curve A. is the mean of three records of the tension of a frog sartorius in Ringer's solution at 0 °C during and after a quick release from maximum tension. Its initial length was $l_0 + 1$ mm.

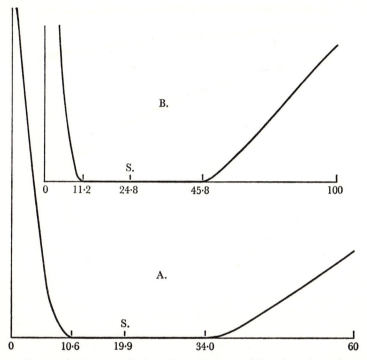

Fig. 4.1. Tension during a rapid release from maximum, followed by shortening under zero tension, then later redevelopment of tension. A., 60 msec sweep; B., 100 msec sweep (only part of release shown). Copied from records like that of Fig. 4.2, but with a higher speed of release. Times to various points given in msec; S., stop. For details and calculation see text.

During stimulation, after 0·4 sec it was released 2·51 mm; at 10·6 msec its tension had fallen sharply to zero; at 19·9 msec (at S.) the release was ended by a stop; at 34·0 msec the tension began to rise again. For a short time after its release (top left) the ergometer accelerated, but over the whole of the significant part of the record its velocity was constant at 150 mm/sec. Between 10·6 and 19·9 msec, therefore, while the tension was zero, the muscle was released

43

a further 1·40 mm (150 mm/sec × 0·0093 sec). Thus between 10·6 and 34·0 msec the muscle shortened under strictly zero tension, and it needed 23·4 msec to take up 1·40 mm slack. Its velocity, therefore, was 60 mm/sec; since l_0 was 25·5 mm, this is 2·36 l_0/sec. The range of length over which this velocity was measured ended at about $l_0 - 1·5$ mm.

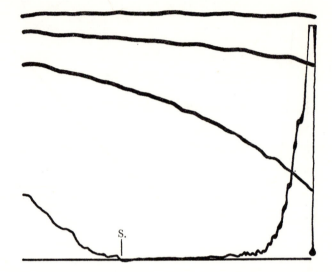

Fig. 4.2. Tension during rapid release from maximum followed by shortening under zero tension, then redevelopment immediately after stop S. Exact copy of record, sweeps 100 msec, read from right to left. The first maximum, before release, was slightly less than the second maximum (shown by final sweep at top) because of the difference of length. Small initial tension shown by spot on right, final tension at the base-line was nearly zero. Release after 0·4 sec stimulus. The velocity of release was equal to that of the fastest fibres, so the falling tension reached the base-line asymptotically, not sharply as in Fig. 4.1, and redevelopment began at once at S.

From 34 msec onwards the tension rose again, at first slowly then more rapidly (the early rise only is shown) and finally reached about the same level as it started at. The gradualness of the early increase of slope can be attributed principally to the time taken by the slower fibres, left further behind between 10·6 and 19·9 msec, in taking up their slack.

In the same experiment (curve B.), at a lower speed of sweep but the same speed of release, the distance released was 3·17 mm. The relevant times are shown in the figure. They allow a speed of 2·32 l_0/sec to be calculated.

In one fortunate experiment (a copy of the original record is in

Fig. 4.2) the speed of release happened to be that of the fastest fibres, which were just keeping up. 'Stop' S. was indicated automatically in the record by a sharp little nick in the curve, and the tension immediately began to rise. The velocity was $2 \cdot 14 \, l_0$/sec.

Various other experiments were made in the same way as that of Fig. 4.1. They agreed in showing that the intrinsic speed of the fastest fibres of the frog sartorius at 0 °C usually lay between $2 \cdot 1$ and $2 \cdot 3 \, l_0$/sec.

After these experiments had been made I learnt that B. M. Millman (1963) had used the same method with the adductor of the oyster. At 21 °C the maximum unloaded velocity was $0 \cdot 24 \, l_0$/sec.

§2. The delayed isometric contraction

The chief difficulty about analysing the behaviour of muscle, during the earlier stages of an isometric (or other) contraction from previous rest, is that the active state requires a finite time to reach its full intensity; in the frog sartorius at 0 °C this is about 50 msec (see ch. 5, §2), several times longer than the latent period. This difficulty can be avoided by the procedure illustrated in Fig. 4.3.

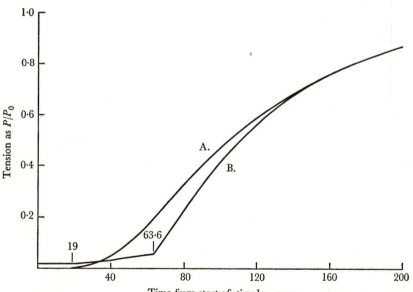

Fig. 4.3. B., delayed isometric contraction, A., ordinary isometric contraction, both at the same muscle length. B. started $1 \cdot 3$ mm longer and was released $1 \cdot 3$ mm, between 19 and $63 \cdot 6$ msec, at 29 mm/sec.

Curve A. is an ordinary isometric contraction at length $l_0 - 1$ mm at 0 °C. The muscle (sartorius of *Rana temporaria*) was on a carrier with multiple electrodes in Ringer's solution. It was stimulated by a maximal induction shock followed immediately by alternating

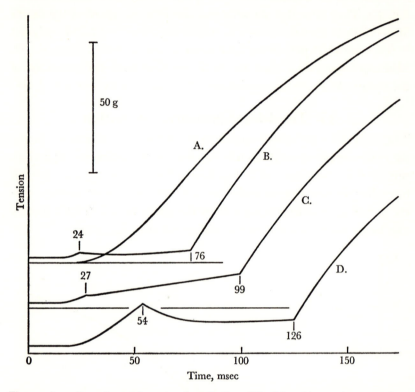

Fig. 4.4. A., ordinary isometric contraction; B., C. and D., delayed isometric contractions, all at the same muscle length. Distances and speeds of release are in the text. The base-lines of C. and D. are lowered to avoid confusion. Times to different points are in msec.

current, 30 c/s, for 0·5 sec. Curve B. is a 'delayed isometric con-traction' made under the same conditions except as follows:

(*a*) initial length $l_0 + 0·3$ mm, at first isometric; then

(*b*) release at 19 msec at 29 mm/sec (0·90 l_0/sec), distance of release 1·3 mm;

(*c*) release ended by a stop at 63·6 msec, contraction now iso-metric at length $l_0 - 1$ mm as for curve A.

The speed of release was low enough for most of the fibres to keep up, so the form of curve B. is concave downwards from the start (the opposite of curve A.). After about 160 msec the two curves run together to the end; *but this is a fortunate chance depending chiefly on the choice of the times of start and end of release.* Usually they run parallel (or nearly parallel) but do not coincide. In Fig. 4.3, contraction B. became isometric at a time by when the active state of the muscle should have been fully developed.

Another experiment is illustrated in Fig. 4.4, with one ordinary isometric contraction A. and three delayed isometric contractions, B., C. and D. The isometric parts of all four were at length $l_0 = 29$ mm. The times of start and end of release are given in msec on the curves. Distances and speeds of release were:

B. 1·86 mm, 1·16 l_0/sec; C. and D. 1·75 mm, 0·84 l_0/sec. As in the experiment of Fig. 4.3 the velocities of release were rather low, and most of the fibres could not have become slack; so the delayed isometric curves are concave downwards throughout.

Fig. 4.5, from another experiment, shows a comparison of the effects of higher and lower velocities of release. Details are given in Table 1. The intention was (p. 48):

Table 1. *Details of the experiments in Fig.* 4.5

No. of record	2.	3.	4.	6.	7.	8.	9.
Time to release (msec)	20·6	20·2	20·2	20·1	20·2	19·3	20·1
Distance released (mm)	3·5	3·2	2·94	2·65	2·32	2·04	1·69
Velocity of release (mm/sec)	76·5	69·3	62·3	55·6	48·7	43·8	38·7
Velocity of release (l_0/sec)	2·59	2·35	2·11	1·88	1·65	1·48	1·31
Time to stop (msec)	66·3	66·4	67·3	67·7	67·9	65·8	63·8
Initial tension (g)	6·3	4·9	4·2	3·9	2·7	2·1	1·6

Notes.

(*a*) times are from the start of the stimulus;

(*b*) P_0 was 95 g; l_0 was 29·5 mm;

(*c*) owing to the high speed for record 2. the tension fell to zero, about 1 g below the final base-line;

(*d*) the records were made in the order of their numbers; they were preceded by 1. (not shown) and followed at the end by 15. (isometric); 5. went wrong and is omitted.

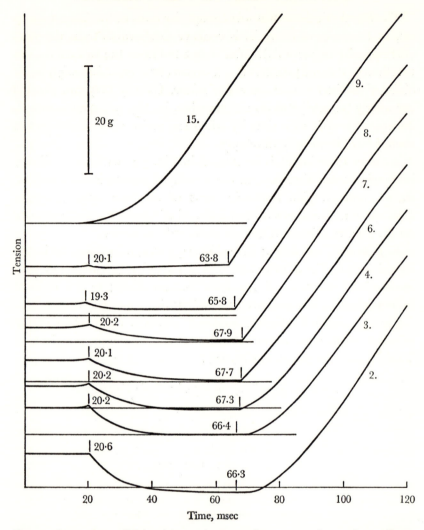

Fig. 4.5. Seven curves of delayed isometric contraction, 2.–9., and one of an ordinary isometric, 15.; all at the same final length, $l_0 - 1$ mm. The speed and distance of release, and other details, are in Table 1. Stimulation began at the vertical line on the left, release began at about 20 msec and ended at about 66 msec, as shown on the curves. For explanation and discussion see text. Tension vertically, 20 g, shown by vertical bar.

(a) That release should end in every contraction at the same length $(l_0 - 1$ mm). This was arranged by keeping the stop fixed;

(b) that release should begin at a constant time-interval after the start of the stimulus, it varied in fact only between 19·3 and 20·6 msec;

(c) that speed and distance should be chosen so as to get the time interval between start and stop as nearly constant as possible; thus the end of release would come at a nearly constant time from the beginning of stimulation.

Since the object was to examine the effects of speed of release over a wide range, it was necessary that the distance travelled, and therefore the initial length, should vary considerably; in the event the initial length varied between $l_0 + 2 \cdot 5$ mm and $l_0 + 0 \cdot 69$ mm. The final length was always the same, $l_0 - 1$ mm.

This was an amusing, and somewhat exacting, experiment to make because one had to calculate between each pair of records (taking account of what had happened already) what should be the magnitude and speed of the next release; then set these on the instruments. It went on longer than shown, but nothing fresh turned up in the later records. The result was better than one might have expected.

The resting tension at length $l_0 - 1$ mm, after the muscle had relaxed, was small, probably less than 1 g, but it was not practical to measure it. The initial tension was measured above the base-line, which was recorded automatically a few seconds after relaxation.

Stimulation began at the vertical line on the left, release started at moments shown by the seven short lines above the curves at about 20 msec, and 'stop' at the short lines at about 66 msec.

Records 2. to 9. were made in that order, 5. is omitted; an isometric record 15. was made later and is shown at the top.

In records 2. and 3., with velocities $2 \cdot 59$ and $2 \cdot 35$ l_0/sec, no rise of tension could be detected, after the stop, before about 4 and 3 msec later. All the fibres had gone slack. In the other records the tension began to rise immediately after stop. Clearly the maximum velocity of the fastest fibres came between curves 3. and 4., somewhere between $2 \cdot 35$ and $2 \cdot 11$ l_0/sec; which agrees with what was found in §1 above. All these curves except 8. and 9. started with concavity facing upwards.

In another experiment the record of Fig. 4.6 was made with a much higher sensitivity for tension. With a release of $3 \cdot 5$ mm at $2 \cdot 44$ l_0/sec the tension dropped to zero and there was a clear gap of about 6 msec after 'stop' before any rise of tension could be detected.

The results of Figs. 4.3–4.6 show that with higher speeds of release the initial concavity of the record after 'stop' faces upwards, with lower speeds it faces downwards, with intermediate speeds the record runs straight out before turning down. I can see no sensible reason for this regular behaviour except that a muscle is made up of fibres with a fairly wide distribution of intrinsic speeds. Some of the fibres of a frog sartorius at o °C can shorten with a velocity around 2·2 l_0/sec. An *average* velocity of shortening of the whole muscle under

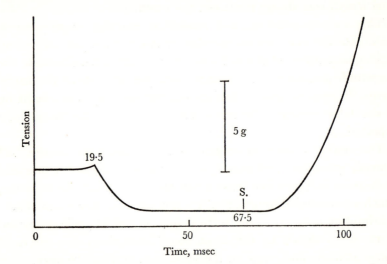

Fig. 4.6. Delayed isometric contraction. Release 3·5 mm at 2·44 l_0/sec, final length $l_0 - 1$ mm. Maximum tension 80 g, 5 g shown by vertical bar. Times in msec from the start of stimulation. S is 'stop'.

zero load can be calculated from the force–velocity relation; it is $P_0 b/a$. With $a/P_0 = 0·25$ and $b = 0·325\ l_0$/sec, this is $1·4\ l_0$/sec. With a maximum of 2·2 l_0/sec and an average of $1·4\ l_0$/sec one could expect a minimum of perhaps 0·8 l_0/sec.

It is easy to see why the curves with the highest velocity of release are concave upwards: most of the fibres get left behind, become slack and join in the redevelopment of tension only as they become taut again. With the lowest velocities of release all the fibres are working together, none becomes slack, and the tension develops along a curve concave downwards. If the myogram is concave upwards after high velocities of release, and concave downwards after low velocities, there must be a range of velocities within which it appears nearly straight.

It would be much nicer if, instead of a general statement of this kind, one could give some sort of quantitative description of what is happening. But, whenever I tried, it became evident that the situation was far too complicated for any exact treatment. The general conclusion, however, is clear, the intrinsic velocities of the constituent fibres of the frog sartorius vary over a fairly wide range. In §1 it was shown that some fibres are much faster than the average; but this could not produce the effects described in §2, these require a fairly wide distribution within an average.

The method of producing a delayed isometric contraction, as described above, is simply to allow the muscle to spend the time it needs to develop full activity in shortening before it finally becomes isometric. In the experiments referred to in §2 this was done by letting it pull against an ergometer moving at a fairly high constant speed so that its tension did not rise or fall much. This is convenient for some purposes, but a simpler method is to allow it to shorten isotonically under a small load for a suitable distance, and then to become isometric. The load, distance and initial length are chosen so that it arrives at the desired final length at a time when it is fully active. This method has been found to work effectively.

§3. General consideration

If the constituent fibres of a muscle can differ considerably from one another in intrinsic speed that could have a bearing on many of its observed mechanical properties. Quicker muscles do nearly everything quicker; their latent period is less, the rise of the active state after a shock is more rapid, probably the decay of the active state starts earlier, the rate of stimulation necessary to give a smooth tetanus is higher; the constant b/l_0 of the force–velocity relation and the rate of heat production during a maintained contraction are greater. Only the characteristics which do not involve any rate or velocity would be expected to be much the same in quicker and slower muscles. Such are, for example, the constant a/P_0 of the force–velocity relation, and the isometric heat coefficient Pl/H in a twitch; these are dimensionless numbers.

The conclusion that a muscle is made up of fibres with a fairly wide distribution of intrinsic speeds adds a statistical complexity to many experiments which a physiologist wants to make. I have, for example, spent much time applying stretches to muscles, in attempts to find out the time course of the development of the active state (see ch. 5). Although the general result is now clear a more accurate description has been thwarted by the presence in muscle of fibres with widely different speeds. Another example of the complication is shown in the results of Ritchie's excellent paper (1954) on 'The duration of the plateau of full activity in frog muscle'. Using a very sensitive method he found that, following the last shock of an isometric tetanus at 0 °C, the onset of decay of tension began about 35 msec later. But 35 msec is much less than the time (say 100 msec) at which the active state in a whole muscle begins visibly to decay after an exploratory stretch; it is even less than the 50 msec required to reach full activity (see ch. 5, §2). But it was easy in Ritchie's records to detect a drop of tension of 1 dyne, which is *about 2% of the maximum tension developed by a single fibre of 50 μ diameter*. The early beginning of relaxation of a single very quick fibre would give his result, and the slower fibres would gradually join in. This illustrates the danger of assuming that all the fibres of a muscle are alike.

Another problem raised by the present results is why the force–velocity relation, as commonly observed, is apparently so simple if the intrinsic speeds of the constituent fibres of the muscle differ widely from one another. This is considered in §4.

§4. The statistical nature of the force–velocity relation in a whole muscle

The usual equation with two adjustable constants fits the force–velocity relation well; no consistent deviations are found except (Fig. 3.2) at very low forces and high velocities, where they were attributed to the presence of a certain number of fibres with high intrinsic speeds. But in §3 it was concluded that a much wider distribution of intrinsic speeds really exists, of which no sign is evident in the force–velocity relation ordinarily observed. This raises a problem to which an answer was sought by using a statistical model, by inventing a possible distribution of the maximum velocities v_0 and testing numerically what sort of force–velocity relation it would give in a composite muscle.

Let the velocity of shortening of the muscle as a whole be v, and assume (for simplicity) that the fibres all have the same length l_0, the same maximum tension P_0, and the same value of a/P_0 ($= 0.25$); but that individual fibres have widely different values of the maximum velocity v_0. When the muscle is shortening with velocity v, the fibres whose $v_0 < v$ become slack and exert no force; but those whose $v_0 > v$ are taut and contribute to the total force.

In order to provide a better test, the statistical distribution chosen was a wide one. It contains 82 fibres with 10 different maximum speeds v_0 as follows:

v_0, as l_0/sec	2.4	2.2	2.0	1.8	1.6	1.4	1.2	1.0	0.8	0.6
n = no. of fibres	1	3	7	13	17	17	13	7	3	1

It is symmetrical and rather like an ordinary probability curve. In a frog sartorius at $0\ ^\circ$C fibre velocities as high as $2.3\ l_0$/sec were described in §1 above, and $1.4\ l_0$/sec is an ordinary mean value for a whole muscle. For the lowest fibre velocities assumed, 0.6 and 0.8 l_0/sec, no direct experimental evidence exists and it might be difficult to obtain it. One reason for introducing these low velocities was to balance the known higher velocities at the other end. There must be a significant number of the fast fibres; otherwise I could not have got (1951) a velocity as high as $1.9\ l_0$/sec in isotonic contractions under 1 g load, with unavoidable friction, viscosity and inertia to be overcome.

The usual force–velocity equation was employed. It was written in the form, suitable for the computation that follows:

$$P/P_0 = \frac{1.25 v_0/4}{v + v_0/4} - 0.25. \qquad (1)$$

If the v_0 of any fibre is greater than the velocity v of the whole muscle, the contribution of that fibre to the total force is given by eqn 1; if its v_0 is equal to v, or less, its contribution is zero.

Now for any chosen value of v calculate, from eqn 1, for each v_0, the value of P/P_0 and multiply it by n, the number of fibres assumed in the table to have that v_0. Add the results together to obtain

$\sum\limits_{v} nP/P_0$. This is the total force exerted by the assumed 82 fibres, slack or taut, when the muscle is shortening with velocity v. Divided by 82 it is the average force. Do this for a number of values of v;

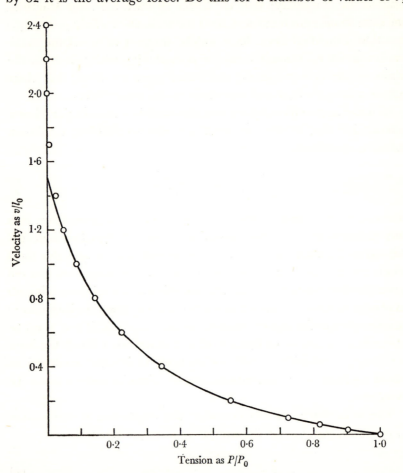

Fig. 4.7. Force–velocity relation, calculated as described in the text for a statistical model. The circles give the calculated velocities at various tensions. The line drawn through the lower ten circles is derived from the usual equation, with constants $a/P_0 = 0{\cdot}23$, $b/l_0 = 0{\cdot}35$.

14 values of v were chosen from $0{\cdot}03$ to $2{\cdot}4$ l_0/sec. Then plot v against $\frac{1}{86} \sum\limits_{v} nP/P_0$. The result is given in the circles in Fig. 4.7.

The curve drawn in the figure was calculated from the ordinary force–velocity equation, but with constants $a/P_0 = 0{\cdot}23$ and

$b/l_0 = 0\cdot35$ $(v_0/l_0 = 1\cdot52)$. These were chosen to give the best fit to the circles at tensions above $P/P_0 = 0\cdot05$. The agreement is rather good and the constants are close to the common assumed value $a/P_0 = 0\cdot25$ and to the average $v_0/l_0 = 1\cdot5$ taken for the statistical distribution. Only at tensions less than $P/P_0 = 0\cdot05$ do the circles diverge appreciably from the calculated curve, but at and near zero tension they are far above it. The appearance of Fig. 4·7 is strikingly similar to that of the experimental results shown in Fig. 3·2.

It is natural to ask whether there is any way, experimental or histological, of recognizing the fibres which have a high, or a low, intrinsic speed. The frog sartorius is made up of fibres of widely different diameters; are the thick ones faster and the thin ones slower? On the analogy of nerve fibres the action potential might have a higher velocity in the thicker muscle fibres, though this would not, in itself, cause more rapid shortening in a muscle stimulated 'all over'. According to Bárány (1967) in a very wide range of muscles the intrinsic speed of shortening varies directly with ATPase activity; and Buller & Mommaerts (1968) have shown that, when a fast-twitch muscle is changed into a slow-twitch muscle by experimental nerve crossing, the ATPase activity is reduced. Will anyone undertake the laborious task of measuring v_0/l_0 and/or ATPase activity, in a number of single fibres in a frog sartorius?

References

Bárány, M. (1967). *J. gen. Physiol.* **50**, 197.
Buller, A. J. & Mommaerts, W. F. H. M. (1969). *J. Physiol.* **201**, 46P.
Hill, A. V. (1949). *Proc. Roy. Soc.* B, **136**, 420.
Hill, A. V. (1951). *Proc. Roy. Soc.* B, **138**, 329.
Millman, B. M. (1963). *J. Physiol.* **173**, 238.
Ritchie, J. M. (1954). *J. Physiol.* **124**, 605.

5

THE DEVELOPMENT AND DURATION OF THE ACTIVE STATE IN STIMULATED MUSCLE, AS REVEALED BY AN APPLIED STRETCH

§1. General considerations and description of methods

As an essential part of the process of developing tension during activity the contractile component has to shorten and stretch the elastic component. The amount of such shortening and stretch depends on the kind of muscle considered, on the sort of contraction, whether twitch or tetanus, and on the compliance of the connexions to the recording apparatus. In the isometric tetanic contraction of a frog sartorius, with inextensible connexions, this shortening is 2–3 % of the muscle's length.

The internal shortening and lengthening greatly slow the development of tension under externally isometric conditions; but by how much it is impossible at present to say. Even if the whole of the series compliance, other than that within the sarcomeres, could miraculously be spirited away, the rest of it, residing in the filaments of the contractile system, might still be substantial (see ch. 7, §7). Under such imaginary conditions one can only guess that tension might develop about twice as fast as normally. The converse effect of increasing, instead of reducing, the series compliance is easy to verify (1951 a).

But even if the series compliance cannot be miraculously removed its effects can be largely reduced by the simple mechanical trick of applying a suitable stretch to a muscle during the early stages of contraction. This saves the contractile component from wasting time in stretching the series compliance to the full tension which the muscle can exert. In fact, the tension of a frog sartorius during tetanic stimulation can thus be raised (§2 below) to its full isometric level in about one-fifth of the time normally required by stimulation alone.

56

And this is a considerable underestimate of how effective the stretch really is; for most of the delay, between start of stimulus and maximum tension, is due to the latent period and to the gradualness of development of the active state. The effects of both could be avoided by using a delayed isometric contraction (ch. 4, §2), instead of one from previous rest. A small quick stretch applied at the start of such a delayed contraction would reduce the time to maximum tension greatly. That is one of the many experiments I might have made and never did; one can only guess that the time to maximum tension in a tetanic contraction of this kind might be reduced from 250 to 20 msec, the former being reckoned from the start of stimulation, the latter from 'stop' of the delaying shortening.

In 1949, twenty-five years after the work with Gasser described in ch. 1, I began trying again, by timed and adjusted stretches, to find out how early, after a single stimulus, the active state of a muscle reaches its full intensity as defined by the tension it can hold. But here, seventeen years later, I came on a serious snag: the frog sartorius (a good—and the traditional—muscle for such experiments) is made up of fibres with different intrinsic speeds (ch. 4). Now it nearly always happens in biology that structures or cells which are quicker in some respects are quicker also in others; one would be astonished to find a muscle fibre possessing a higher intrinsic speed which did not also have a shorter latent period, or develop its active state more quickly, or begin to relax earlier, or need a higher frequency of stimulation to give a smooth contraction. At any rate let us *assume* that in the frog sartorius the fibres of higher intrinsic speed develop their active state in a shorter time, and their maximum velocity earlier. Let us further assume that, after a single stimulus, the intensity of the active state, *as measured by the force the muscle can just hold*, rises and later falls along a curve like that marked P_i in Fig. 5.1 A. And finally let us assume that the time-scale of curve P_i is different in slower and quicker fibres, the cycle being completed in a shorter time in the quicker ones.

Now in a muscle with a uniform population of fibres, in order to get its full tension developed as early as possible, its elastic component would need to be stretched at such a rate that its tension followed the rising limb of curve P_i. If such a uniform muscle could be realized it would not be difficult, by trial and error, to find out, approximately at least, the best form of such a stretch. That is what Gasser and I, in the original study of this subject, tried to do in a primitive way in 1924 (see ch. 1). We concluded

The fundamental mechanical response is shown to attain its maximum intensity quite early in the contraction, long before the maximum tension is developed, and to pass off continuously thereafter.

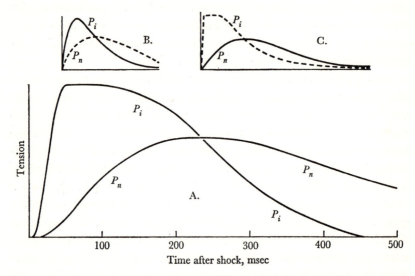

Fig. 5.1. A., P_i is the supposed time-course of the 'intrinsic strength' of the contractile component of a muscle after a single shock; it is defined as the tension under which the muscle would neither shorten nor lengthen. P_n is the tension observed in a normal twitch. B., the same idea in 1924 (Gasser & Hill); C., in 1949 (Hill).

The last sentence of course referred to a twitch. In a tetanic contraction the 'maximum intensity' is maintained by each of a succession of stimuli. Fig. 5.1 B., taken from Fig. 16 of our paper (ch. 1 above), shows what we then imagined the fundamental response looked like.

In reporting in 1949 the results of the second attack on the problem, I made the following claim:

There is strong converging evidence that the change of state from rest to full activity occurs abruptly very early after a stimulus.

58

This was an exaggeration, which found a place in the literature (as exaggerations are apt to do) and needs correction; *the change does occur early, but not abruptly.* Its time-course was not easy to determine, partly because of imperfect methods, partly because of the composite nature of the muscle. Fig. 5.1 C. curve P_i represents what I imagined, in 1949, was the shape of the fundamental response.

I had long thought that the experiments described in the 1949 paper ought to be repeated more critically and with better methods, so I started out again in the autumn of 1964. Much time was spent in trying to get more clear-cut results; and it was only in the autumn of 1966 that the nature of the difficulty that had dogged one's footsteps was realized—from the results of quite different experiments (see ch. 4, §3).

Consider what happens in a fast group of fibres, during rising activity, when it is stretched at a rate suitable to match the development of activity in a slow group. It reaches its maximum tension before the stretch stops, its own shortening adding to the effect of the stretch; then, continuing to be stretched, its tension rises still further and finally drops back quickly as soon as the stretch ends. Now, conversely, consider what happens when a slow group of fibres is stretched along a curve P_i which matches the development of activity in fast ones. The tension rises quicker throughout than its fibres can tolerate, and it 'gives' irreversibly throughout the stretch. Finally, when the stretch ends the tension drops back. These effects can be seen in some of the curves of Figs. 5.8 and 5.9.

If the stretch followed a curve P_i, matching the development of activity in the main group of fibres of moderate speed, the faster and the slower fibres would both be overstretched in the way just described. The tension would overshoot its final level, then drop back quickly and gradually redevelop.

It is impossible, therefore, in a composite muscle, to do the simple experiment of finding out by stretching when the active state is fully developed. Various compromises had to be made and the results interpreted. The most recent experiments were performed before the nature of the difficulty was realized; their results can now be interpreted better, and an account of the most significant results is given in §§2 and 3.

Several things worried me about the 1949 paper:

(*a*) the stretches used were excessive, 10–15% of l_0;

(*b*) there was no statement of the compliance of the connecting wires and recorder, these may have been large. If so, *net* stretches may not have been so excessive;

(*c*) the stretches were at constant speed and quite unsuitable for anticipating the shortening, at a decreasing rate, of the contractile component as its tension rose;

(*d*) there was always a considerable overshoot in the records. This was due to overstretching the contractile component, and the excess of tension so developed was rapidly reversed when the stretch stopped.

In order to avoid the overshoot just mentioned, and to make the stretch more nearly match the time-course of rising activity during stimulation, linear stretches were avoided. Instead, in all the significant experiments made in the latest investigation (October 1964–May 1966) the stretch (*s*) was either purely exponential,

$$s = pt_0 \left(1 - e^{-(t-t_1)/t_0}\right) \quad [t = t_1 \text{ to } \infty],$$

or, more usually, linear followed by exponential,

$$s = p(t-t_1) \; [t = t_1 \text{ to } t_2] + pt_0(1 - e^{-(t-t_2)/t_0}) \; [t = t_2 \text{ to } \infty].$$

Here t_1 (msec) is the time after the stimulus at which stretch began, t_2 that at which the linear stretch ended, t_0 the time-constant of the subsequent exponential and p (mm/msec) the intial rate of stretch. The stretches were made with the ergometer shown in Fig. 1·2, p. 134, of my book (1965). It has a long strong spring which by itself would give a linear stretch. But if, or as soon as, the motion of the ergometer arm was opposed by a short spring the stretch became exponential. There was no discontinuity of velocity when the short spring came into action, only a deceleration. Various short springs of suitable strength and length were constructed and the most appropriate chosen for any task.

The speed p was adjustable with a calibrated setting and the total stretch pt_0 in the exponential part could be measured; so, p being known, t_0 could be calculated. The movement of the ergometer was ended not suddenly by a stop (which causes vibration) but gradually by the spring. In this way the sharp overshoot of tension at the end of stretch was largely, or completely, avoided. The records in §§2 and 3 show how much better this arrangement proved to be than that employed in 1949.

The simplest case investigated (§2 below) was that of a tetanic contraction, in which (if the tetanus was long) only the development of the active state and not its decay had to be considered. It was necessary of course to know the exact moment of the first shock of the stimulus; that was secured by using a pair of rapid relays which simultaneously started the sweep of a cathode ray tube and broke the primary circuit of a good old-fashioned induction coil providing a supermaximal shock. The rest of the stimulus was supplied simply by an alternating current of suitable low frequency, usually 15–25 c/s, though sometimes 50 c/s when the stimulus was very short. The alternating current began to flow immediately after the induction shock was fired. The ergometer was released by breaking the circuit of its electromagnet at any desired interval after the stimulus started. The lag in starting up was very short because the driving spring and the damping were strong.

A muscle carrier with multiple electrodes was used, to ensure nearly simultaneous excitation all over; this, with the muscle on it, was in oxygenated Ringer's solution at 0 °C. Tensions were recorded by an R.C.A. transducer (5734) carried on the ergometer arm, and in the later experiments (all those in §2 and a few in §3) the connector consisted of four strands of stainless steel wire (100 μ) twisted together, annealed under tension to straighten and treated with 'araldite' or 'yacht varnish' to bind them. The compliance of the connector (which had to be rather long) was usually about one-third of that of the series elastic component of the muscle. It was always calibrated. In most of the experiments of §3 (which were made before those of §2) the connectors consisted of a similar construction of braided nylon or terylene thread. These were more extensible than the steel wires used later, but they were calibrated and their imperfections did not invalidate the results.

The other case (§3) to be investigated was that of the rise of the active state after a single shock and its later decay. The earlier investigations of 1924 and 1949 had been concerned with this alone.

61

The methods used were the same as those just described, except alternating current was not required and the connecting thread had a greater compliance. For most of the experiments referred to in §3 toad sartorii were used, chiefly from a single batch of large 'continental' toads[1] (*Bufo bufo*) which remained in excellent condition for months. A few experiments were on the much smaller English toads, others on English frogs.

In both series of experiments a record was always made of an ordinary isometric contraction at the length to which the muscle was stretched in the other contractions.

§2. The effect of a controlled stretch applied early during a tetanus

The original purpose of the investigation to be described in this section and the next was to re-examine in general the findings of my 1949 paper. But attention was soon concentrated on three special questions:

(1) Is it true that by the application of a suitable stretch the tension of a stimulated muscle can be raised rather quickly to a level (the 'plateau') at which it remains constant for some time after the stretch ends?

(2) How soon after the start of stimulation can that level be reached? and

(3) How long does the plateau last after stretch and stimulus are over?

This section deals only with tetanic stimulation, usually quite short; the following one with twitches in response to single shocks (this was the only case considered in the 1949 paper).

The experiments of §2 were made on sartorii of English *Rana temporaria*, some before the breeding season, some after. In every experiment the following conditions could be varied, in order to get the most significant results: (*a*) the time after the start of stimulation when stretch began; (*b*) the amplitude of the stretch, and whether exponential, or linear followed by exponential and in what proportion; and (*c*) its speed. The choice of any of these influenced the

---

[1] The dealer who supplied them informs me that *Bufo bufo* is found throughout Europe, except in the Mediterranean areas; the animals I used were merely larger and better specimens than the native variety.

choice of the others, so a good deal of trial and error usually occurred before the best result was obtained. The experiments described below were themselves typical; but they were more fortunate than many others in the fact that the right conditions were found before the muscle showed obvious signs of failure. This was not the sort of investigation in which an average result, with its standard deviation and all, was much use. With so many variables involved the question, rather, was whether one could hit the target or not, and if so where.

In the figures the start of a stretch is shown by a short vertical bar, the end of the linear part of the stretch by ↑, the effective end of the exponential part by ↓ ('effective' meaning within $e^{-3} = 5\%$, which with a 0·25 mm spring is 12·5 μ). In cases where the tension was still rising at the right hand end of a figure, the later maximum is shown by a horizontal bar and the letter M. The end of stimulation is shown by a filled circle.

Experiments

Fig. 5.2 illustrates an experiment in which the start of the stretch was very early, well before the end of the latent period. The stimulus lasted only 60 msec. Details are in the legend. The records were made in the order 1., 2., 3. Of these, 1. suggested that a faster stretch would be better, but 2. was then evidently a bit too fast, so 3. was made at an intermediate speed. That was rather successful, the plateau in 3. was reached in 50 msec and lasted for more than 100 msec. The early start of the stretch had the consequence that much of it was 'wasted' in stretching fibres only just beginning to be active. So long stretches were needed.

Fig. 5.3 contrasts with Fig. 5.2 in the fact that the stretch started later, so could be much smaller; in fact it was still very slightly too large as shown by the gradual small decline of tension after the end of the stretch.

Fig. 5.4 gives another example of a small stretch following which the tension remained practically constant for another 150 msec after the end of the stretch (and of the stimulus).

In Fig. 5.5 (made with the same muscle and stimulus as Fig. 5.2) the interest resides in two successive records in the series, 9. and 10. In 9. the total stretch was 2·2 mm and the slight rise of tension after

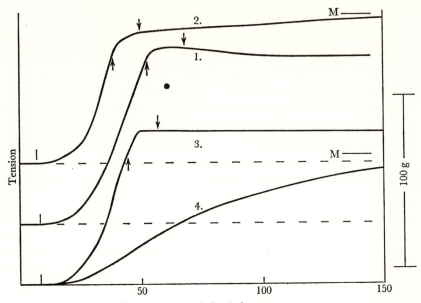

Time from start of stimulation, msec

Fig. 5.2. The effect of an early controlled stretch during a short tetanus (0·06 sec). Frog sartorius, 0 °C, $l_0 = 31$ mm, $M = 125$ mg. *Three upper curves*, stretch 2·25 mm linear, 0·25 mm exponential; starting, from above downwards, at 6·6 msec, 8·0 msec and 8·2 msec respectively; speeds 67, 49, and 59 mm/sec, t_0 3·7, 5·0 and 4·2 msec. *Bottom curve*, isometric 'long', maximum tension at about 240 msec. Vertical bar at right 100 g.

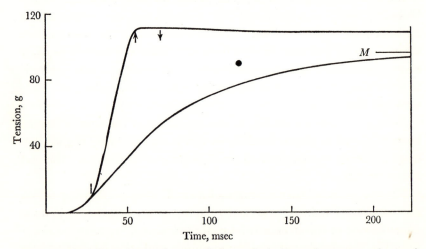

Time, msec

Fig. 5.3. The effect of an early controlled stretch during a short tetanus (0·12 sec). Sartorius, 0 °C, $l_0 = 31$ mm, $M = 135$ mg. *Upper curve*, stretch 1·25 mm linear, 0·25 mm exponential, begun at 27·7 msec from $l_0 - 1$ mm, 45 mm/sec, $t_0 = 5·6$ msec. *Lower curve*, isometric at $l_0 + 0·5$ mm; M, maximum tension at about 300 msec.

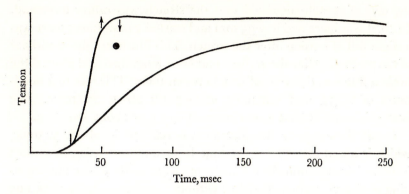

Fig. 5.4. The effect of an early controlled stretch during a short tetanus (0·06 sec). Sartorius, 0 °C, $l_0 = 28·5$ mm, 92 mg. *Upper curve* stretch 1·25 mm linear, 0·25 mm exponential, began at 28 msec from $l_0 + 2$ mm, 59 mm/sec, $t_0 = 4·2$ msec, maximum 72 g. *Lower curve* isometric at $l_0 + 3·5$ mm, maximum 60 g.

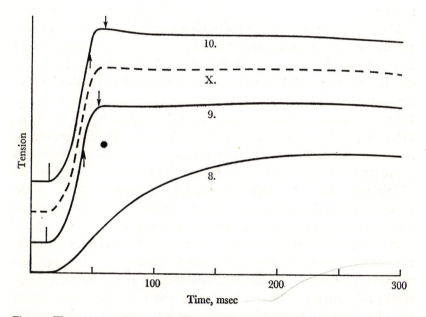

Fig. 5.5. The same experiment as for Fig. 5.2. No. 8., isometric contraction at $l_0 + 4$ mm recorded just before no. 9. No. 9., stretch begun at 12·5 msec, 1·95 mm linear, 0·25 mm exponential, at 61 mm/sec, $t_0 = 4·1$ msec. No. 10. stretch begun at 14·3 msec, 2·1 mm linear, 0·25 mm exponential, same speed. The broken curve X is drawn half way between 9. and 10.; see text. Maximum for 8., 69 g; for 9., 81 g; for 10., 85 g.

the end of stretch suggested that the stretch was rather too small. In 10. it was increased to 2·35 mm and started 2 msec later; there was a slight fall of tension after the stretch. This illustrates the sensitivity of the method. After the records had been magnified and plotted the broken line was drawn half way between them. This rose to a maximum of 83·5 g and remained at 83 g till about 250 msec. The isometric curve 8. had a maximum of 69·5 g at about 260 msec.

In all these figures the plateau was reached in about 50 msec, which is about one-fifth of the time to maximum tension in a short isometric contraction. Many attempts were made to get the beginning of the plateau earlier, but without success whatever trick one employed. It was easy of course by a quicker stretch to make the tension reach much earlier, or exceed, what would later prove to be its final level; but then the tension collapsed and only gradually reached its final level as the contractile component continued to shorten.

If the curves of Figs. 5.2–5.5 can be assumed to show that most of the fibres of the muscle reach full activity in about 50 msec, it can be concluded that some of them must reach it in considerably earlier, say, in 25 msec; and that is about the moment at which the velocity of shortening under a very small load attains its maximum (1951 *b*). It has been doubted (Huxley 1957, p. 293) whether this early attainment of maximum velocity of shortening can be interpreted as meaning that activation is then complete. Certainly it does not prove it; but in a population of fibres with considerable differences of intrinsic speed it seems pretty certain that some fibres would become fully active earlier than others, *and these are the ones that would determine the velocity of shortening under a very small load.*

In all experiments of this kind it was found that the tension during the plateau produced by a stretch was substantially higher than the maximum isometric tension of a contraction at the same (i.e. the stretched) length, with a stimulus of the same duration. Many experiments were made with a stimulus long enough to give the full isometric tension, and in these the plateau was, on the average, 14% higher. The difference never failed to appear. It might be thought that the excess of tension after stretch, though lasting for a long time, could be due to the machinery getting 'locked' in a strained condition from which it would gradually relax. There is no evidence for this; the muscle does not relax till after the stimulus ends, and then normally. The experiment illustrated in Fig. 5.6 shows that after a stretch was ended a muscle could still shorten slowly against a tension substantially greater than it could develop under isometric conditions.

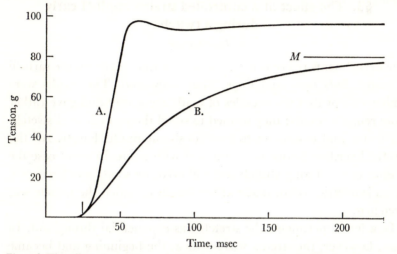

Fig. 5.6. The effect of an early controlled stretch during a tetanus (shock and a/c 15 c/s, 0·6 sec). Sartorius, 0 °C, $l_0 = 29$ mm, $M = 161$ mg. Stretch A starting at 24 msec, 1·71 mm linear, 0·25 mm exponential; isometric B. at the final stretched length, maximum reached by 320 msec.

The effect of a stretch in enhancing (potentiating is the fashionable word) the strength of a muscle, if it occurs also in human muscle, might have a beneficial application to rapid skilled movements of various kinds. Does a sudden quick extension, or jerk, make a muscle capable of holding a greater load and so possibly of supplying stronger support for forcible movements elsewhere? The complicated and powerful movements of skilled performers might be found to make good use of the higher tension developed during a stretch, and maintained afterwards.

Since the last paragraph was written I have seen a paper by Cavagna, Dusman & Margaria (1968) on 'positive work done by a previously stretched muscle'. Experiments were made on toad sar-

5-2

torii, frog gastrocnemii and forearm flexors of man. Stretched during contraction, and allowed to shorten immediately after, a muscle was able to do more work than in shortening at the same speed from a previously isometric contraction. This may be related to the effect described above.

§3. The effect of a controlled stretch applied early during a twitch

A. *Introduction*

The experiments referred to in this section were made on the sartorii of toads (*Bufo bufo*) and frogs (*Rana temporaria*). The results were similar except that the muscles of toads are slower and gave rather more reliable results; they seemed to stand the procedure of stretching better and continued to give consistent results longer. All the results described below were obtained with toads. In every case the tension curve during stretch was followed, or sometimes preceded, by an isometric record made at the length to which the muscle was stretched.

In a few experiments the stretch was exponential throughout. In most, however, the stretch was linear at the beginning and became exponential later. Records were made, usually in two or three successive sweeps, covering also the early decline of tension after the active state had started to decay. As with the tetanic contractions studied in §2 it was necessary to make a good many trials, in order to find the best conditions; the use of single twitches, and of toad muscles, made it easier to get consistent results during a long experiment.

The toads used (*Bufo bufo*) were of two varieties, common English toads which are small, and animals described as 'continental'. The latter were bigger and stronger, but look and behave otherwise exactly like the English ones. See the note at the end of §1.

B. *Experiments*

In Fig. 5.7, A. is for an experiment on an English toad, B. for one on a 'continental' toad. The curves are exact copies of the original records, of tension vertically against time horizontally; they were

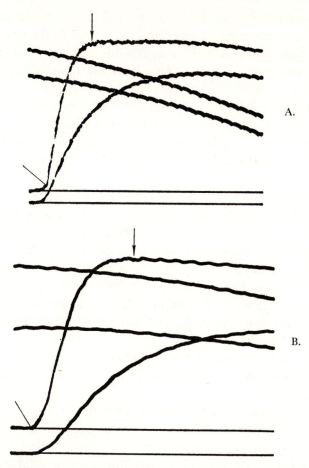

Fig. 5.7. Records of stretched and isometric twitches (o °C) of toad sartorii, *Bufo bufo*, A., English, B., 'continental'. Stretched above, isometric below, each two sweeps. Start of stretch at tip of early sloping line, 95 % complete at arrow. Stimulus at origin.

	l_0 (mm)	Stretch (mm)	Started (msec)	t_0 (msec)	Sweep (sec)
A.	27·5	I	38	32	0·5
B.	32·5	3	24	38	0·3

The stretch in B. started earlier so had to be greater.

made with a cathode ray tube, each with two sweeps, A. of 500 msec, B. of 300 msec. The sweeps started directly the stimulus, a single shock, was applied. In both pictures the upper curve is for a contraction with an early exponential stretch, the lower curve for an iso-

69

metric contraction at the length to which the muscle was finally stretched for the upper curve. In A. and B. alike the stretch began at the time to which a short line points near the origin; and the stretch was 95% complete $(1-e^{-3})$ at the time indicated by the arrow.

In A. the stretch was of 1 mm only $(l_0-1\cdot5$ mm to $l_0-0\cdot5$ mm) with a time constant of 32 msec. The plateau was reached at about 120 msec from the stimulus and the tension began to decline at about 350 msec; the maximum tension in the isometric curve was at about 500 msec.

In B. the stretch was of 3 mm $(l_0-4\cdot5$ mm to $l_0-1\cdot5$ mm) with a time constant of 38 msec. Strictly regarded there was no plateau after the arrow but a slow decline. The stretch was slightly too great and the muscle could not hold the tension developed.

In A. and B. alike the plateau was reached rather later than in Figs 5.8 and 5.9. This was due to the stretch being purely exponential. For the later figures it was linear at first, finishing with exponential, which saves time. As in §2, the first arrow shows the end of the linear stretch, the second the 'effective' end of the exponential one.

The rest of the experiments described in this section were with 'continental' toads, and the records were magnified in a lantern and drawn on squared paper.

Fig. 5.8 shows the sort of procedure that was followed sometimes to obtain the best conditions. The stretch was nominally 3·4 mm, of which 3·05 mm was linear, 0·35 mm exponential $(l_0 = 32$ mm); but the connector was rather extensible, and the total extension of the muscle itself at the maximum of record 5. was about 2·6 mm. The records were made in the order shown, all from the same initial length, with stretch starting (at the vertical line) 28 msec after the shock. The speed of linear stretch varied widely, being in 1. to 7. successively (in mm/sec) 135, 96, 69, 53, 41, 31 and 26. The speed of the final exponential stretch varied in the same proportion. With 1. and 7. also are shown isometric contractions recorded immediately after them; these two are practically of the same height (99 g) showing that the muscle had not changed during the series.

The stretch in 1. was evidently much too fast, the contractile component 'gave' quickly and then after a pause continued to

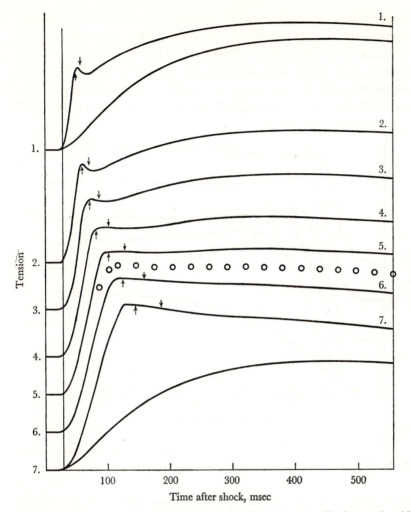

Fig. 5.8. Successive twitches numbered in order from the start. Toad sartorii, 0 °C. Circles are interpolated half-way between records 5. and 6. Isometric contractions followed 1. and 7. at the stretched length. See text.

shorten; the stretch in 7. was too slow, the muscle was overstretched and then gradually lengthened from 125 msec onwards. In 5., however, the tension rose rapidly to reach 130 g at 100 msec, and remained nearly constant for a long time. A stretch intermediate in speed between 5. and 6. would have been best, as shown by the spots interpolated half way between them. A line through these spots

71

would begin to turn down gently at about 350 msec. The isometric curves 1. and 7. gave maximum tension at about 450 msec.

Fig. 5.9 is similar to Fig. 5.8, records 6., 7., 8., 9. and 10. being made in that order, with an isometric record at length l_0 at the bottom. The stretch was nominally 3·3 mm (3·0 mm linear, 0·3 mm exponential) of which (at maximum tension) 0·7 mm was taken up

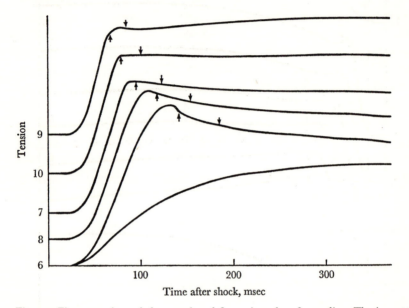

Fig. 5.9. Five successive twitches numbered 6.–10. in order of recording. The isometric contraction at the bottom followed record 7., at the stretched length (l_0). Notice particularly record 10.

by the stretch of the connector. The speeds in their linear parts were, from 10. downwards, 59, 47, 36, 28 and 23 mm/sec, and proportionally in the exponential parts. Stretches all began at 23 msec, before the end of the latent period. Record 10. reached 103 g at 80 msec and remained near there till about 330 msec. The maximum of the isometric contraction (88 g) was at about 380 msec. One reason why records 7., 9. and 10. gave full tension so early was that the stretch started rather before the end of the latent period. Record 10. looks almost too good to be true, but it really did happen. By taking a series of this kind and making a record intermediate in speed between two diverging in opposite directions, one can sometimes obtain a rather superior result; 10. came from bracketing between 7. and 9. But this cannot be done unless the muscle continues to give constant results; so single twitches are better than tetanic contractions and toads than frogs.[1]

Experience has shown that the earliest time at which the plateau is reached, as in Fig. 5.9 curve 10., is best achieved, not by a purely exponential stretch as in Fig. 5.7, but by a combination of a fairly long linear stretch followed by a shorter exponential one; also it is better to start the stretch rather early, not later than the end of the latent period (as normally observed). With such precautions the time at which the plateau reached its full height was about one fifth of the time to the maximum in an isometric twitch.

The *duration* of the plateau of the active state is important. In Fig. 5.8 curve 5. and Fig. 5.9 curve 10., as well as in the original records of Fig. 5.7, it is evident that this is less, but not much less, than the time to maximum tension in an isometric twitch. An attempt was made to get a more accurate estimate. In fourteen separate experiments an obvious plateau was reached. In them the time after the stimulus at which (in magnified records like those of Figs. 5.8 and 5.9) the tension first began visibly to diminish was measured and compared with the time to the maximum of the isometric twitch. The former averaged 66% of the latter, with a S.E. of the mean of 2·5%. On the average, therefore, the end of the active state so measured occurred at about two thirds of the time to the maximum of the isometric twitch.[2] On the average also the time to the beginning of the plateau was about 20% of the time to maximum isometric tension. Of course, partial activity remains much longer.

[1] These records with toads were all made before it was realized how a light connecting wire with very small compliance, and of sufficient length, can be constructed. The fact that the compliance in these experiments was rather large had the effect that the 'linear' stretches were not strictly linear, and the speeds and time constants quoted are not quite exact; these refer to the top of the connector where it joined the recorder, not to the bottom where it joined the muscle. But this does not in any way affect the general validity of the results.

[2] The time to maximum tension in an isometric twitch is determined by a balance between continuing shortening of slower fibres and early lengthening of faster ones.

C. *Discussion*

(1) It is not known whether, in the sartorii of toads, the intrinsic speeds of the constituent fibres vary as widely from one another as they appear to do in frogs. This would be quite easy to verify, but it has never been done. If a toad sartorius is more uniform in this respect, that would be a valuable property for many investigations; it might in fact be the reason why the present study gave clearer results with toads than with frogs.

(2) Most of the methods, electrical or mechanical, of investigating the time taken by a muscle in reaching full activity have assumed that all its fibres are similar in their time relations. This assumption must be regarded very critically, see particularly ch. 4, § 3.

(3) The decline of full activity (the decay of the active state) is bound to be more difficult to investigate in a muscle with a wide dispersion of fibre speeds.

(4) For reasons given in § 1 above, the attempt to reach a plateau of tension early is certain to lead to a stretch of the contractile component of many fibres. The question naturally arises—can this stretch, of itself, cause a lengthening of the active period? There is no evidence that it does—but equally none that it does not. In the case of stretch during a tetanus (§ 2) the tension always rose appreciably higher than in an isometric contraction, and held it for a long time; if anything like a stimulus had resulted from the stretch one might have expected something to remind one of an ordinary response. So also with the stretches of Figs. 5.7–5.9; there is nothing in these to suggest anything like a physiological response to a stretch.

(5) The time-course of the decay of the active state cannot be investigated by stretches; but it can be by releases. The principle is to release a muscle at various times during a twitch, to record the redevelopment of tension, and to measure the maximum redeveloped tension P and the time t at which it occurs. If P is plotted against t the resulting curve is assumed to represent the time-course of the 'intensity' of the active state at times after the maximum of the isometric curve. The assumption is a reasonable one, and is based on the diagrams in Fig. 5.1. It has been used in various investigations, e.g. by Ritchie (1954) and by Ritchie & Wilkie (1955, 1958).

74

It does not give the earlier part of the curve, before the maximum of the isometric twitch; and a better method would be to use a short tetanus (say 0·3 sec, frog, 0 °C) in which the intensity of the active state would be at its maximum long before the end of the stimulus. Then by release and recorded redevelopment of tension the whole course could be mapped out. The release could be before, or after, the last shock, in order to give points all along the curve. Indeed, since maximum tension developed depends on length, it would be better to put the first release at a sufficient time *before* the last shock in order to allow the full tension to be developed at the shorter length, and then to continue with increasing intervals until no tension was redeveloped. *Will someone please make the experiment in this way?*

References

Cavagna, G. A., Dusman, B. & Margaria, R. (1968). *J. appl. Physiol.* **24**, 21.

Gasser, H. S. & Hill, A. V. (1924). *Proc. Roy. Soc.* B, **96**, 398.

Hill, A. V. (1949). *Proc. Roy. Soc.* B, **136**, 399.

Hill, A. V. (1951 *a*). *Proc. Roy. Soc.* B, **138**, 325.

Hill, A. V. (1951 *b*). *Proc. Roy. Soc.* B, **138**, 329.

Hill, A. V. (1965). *Trails and Trials in Physiology*. London: Edward Arnold.

Huxley, A. F. (1957). *Prog. Biophys.* **7**, 257.

Ritchie, J. M. (1954). *J. Physiol.* **126**, 155.

Ritchie, J. M. & Wilkie, D. R. (1955). *J. Physiol.* **130**, 488.

Ritchie, J. M. & Wilkie, D. R. (1958). *J. Physiol.* **143**, 104.

6

THE ELASTIC ENERGY IN A CONTRACTING MUSCLE DUE TO ITS DEVELOPED TENSION

§1. The first fifty years

This, in its origin, is a very old story, so it is introduced by a very old experiment, one made by Levin & Wyman (1927) with their ergometer. They allowed me to describe it in a Croonian Lecture (1926) to the Royal Society. Its results are given in Figs. 6.1 and 6.2

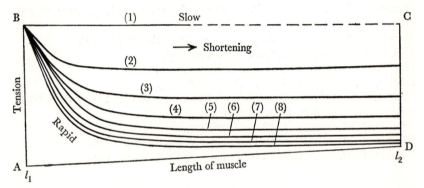

Fig. 6.1. Tension–length curves of tortoise muscle shortening, during tetanus, at various constant speeds. Muscle stimulated isometrically at length l_1 till maximum tension AB developed; then, with stimulus continuing, allowed to shorten at constant speed to length l_2. Isometric tension at $l_2 =$ DC. Speeds, relative values, (1)–(8): 0·083, 0·59, 1·31, 2·27, 3·79, 5·0, 6·6, 10·8 (Levin & Wyman).

here. If anyone reads their paper let him look at the beautiful records of their experiments and disregard the viscous-elastic theory; that was my fault, not theirs.

A tortoise muscle was connected to an ergometer and stimulated at length l_1 to give an isometric tetanus. When the tension had reached its maximum (AB) the ergometer was released at various constant speeds, from very slow to fast, until it reached a stop at muscle length l_2. Eight records (Fig. 6.1) were thus made of tension as a function of length during shortening. The area of any one of these above the base-line is the work done by the muscle in shortening from l_1 to l_2. The work is greater the slower the shortening. At the lowest speed (top) the change of length was nearly 'reversible' in the thermodynamic sense, but stimulation was ended to avoid fatiguing the muscle. The isometric tension at length l_2 was measured

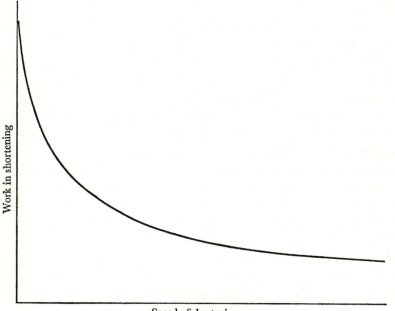

Fig. 6.2. Work done in shortening, calculated from areas of Fig. 6.1, as a function of speed (Levin & Wyman).

(DC) in a separate contraction and the top record was completed by the broken line.

The work performed in the eight contractions was calculated from the area below the curves and is shown as a function of the speed of shortening in Fig. 6.2. A substantial amount of work was done even at the highest speed. This was due to the undamped elastic element in series with the contractile component of the muscle. No doubt some of it resided also in the connector and the tension recorder; but Levin & Wyman were accomplished experimenters (Wyman still is) and would not have allowed a gross error to affect their results. In any case, many years later I made a lot of experiments of this kind with nearly inextensible connexions and found, as they did, that however great the speed of controlled release a finite amount of work was done. This irreducible work is regarded as the elastic energy present in the contracting muscle (and its connector) when under its developed tension.

The contractile component of a muscle shortens at a determinate speed under zero load (ch. 4, §1). Released at any higher speed the contractile component becomes slack. At a lower speed it can do physiological work. In the slower shortenings (upper curves) of Fig. 6.1, the contractile component was doing a good deal of work, in the more rapid shortenings (lower) very little.

In the earliest experiments (Hartree & Hill, 1920) on the time-course of the production of heat by an isometrically contracting muscle it became evident that a substantial amount of the heat appeared while mechanical relaxation was going on. The 'hump' in the records of relaxation heat was attributed to the degradation into heat of the elastic energy developed during contraction. (In the earlier days, before methods were perfected, a substantial part of the elastic energy was really in the connecting wire and tension recorder.) But it was thirty years before anyone tried to measure how great this developed elastic energy is. Then, by a method really analogous to that of Fig. 6.1 (recording the tension–length curve during a very rapid release from maximum tension), the elastic work was determined experimentally on frog muscles (1950). This was followed (1953*a*) by rather more accurate experiments on toad muscle; and (1958) by experiments of the same kind by Jewell & Wilkie on frog muscle.

But already by 1953 the chief interest of these experiments involving a quick release from maximum tension had changed course, and we were coming to suppose that the force–length curve so obtained was the tension–extension curve of that will-o'-the-wisp, the series elastic component. So it is in some circumstances, but more of that in ch. 7.

§2. Eleven experiments and a tension–extension curve

During the summer of 1966 I made eleven experiments, with quick release from maximum isometric tension, on the sartorii of a remarkably consistent batch of English frogs, *Rana temporaria*. The muscles, in Ringer's solution at 0 °C, were connected to a Levin–Wyman ergometer (now recording electrically) and excited by a maximal tetanic stimulus at length about $l_0 + 1$ mm. Then, from maximum tension (the stimulus continuing) they were rapidly released about 1·5 mm at a high velocity (about $4 l_0$/sec, which is roughly twice the intrinsic speed of their fastest fibres) and a tension–time record was made electrically from maximum to zero tension. A calibration curve of the movement of the ergometer was made under similar conditions, which allowed the required tension–distance curve of the muscle to be calculated.

Then a correction was necessary for the shortening of the contractile component during the release, as tension fell. The correction is small but not negligible; and it cannot be very accurate because the slower fibres were not contributing their full quota to the tension, none at all in the last stages of the release. The need for it could be reduced by using a much higher velocity of release; but that would cause more trouble than it was worth, in vibrational disturbance of the records and in hysteresis in the experimental material. Better to bear the ills we have . . . Then a final correction had to be made for the compliance of the connecting wire.

In each experiment the results were expressed by a curve relating tension vertically to distance of release horizontally. Since an average was to be calculated for muscles of different size, tension was expressed in terms of Pl_0/M, distance of release as a submultiple of l_0. In each of the eleven curves the distance of release was read off between $Pl_0/M = 2{,}000$ and various lower tensions. For each of these lower tensions the average value of the distance of release to it was calculated. These averages are given by the dots in Fig. 6.3.

In order to get a measure of the consistency of the results the deviation of the distance of release from $Pl_0/M = 2{,}000$ to six lower tensions was calculated. The horizontal bars above these points give the standard error (S.E.) of the mean ($+$ to $-$). The consistency of the method (and of the frogs) was good.

The eleven experiments on which Fig. 6.3 is based did not allow a direct calculation of the mean curve at tensions above $Pl_0/M = 2,000$. Nevertheless, it was possible to extrapolate the indivi-

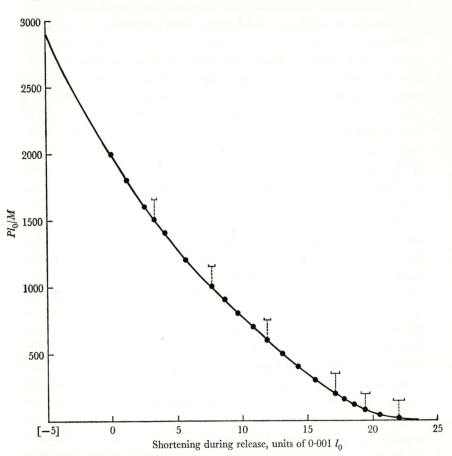

Fig. 6.3. Tension–distance curve of frog sartorius obtained during rapid release from maximum developed tension as described in the text. Ordinate, tension in the generalized form Pl_0/M, abscissa as a fraction of l_0. Mean for eleven muscles. The circles are average values calculated as described in the text, and the six horizontal bars are S.E.'s (+ to −) of the mean distance shortened to the tensions indicated.

dual curves with reasonable confidence, since in this region they were nearly straight, and this allowed the extension of the mean curve of Fig. 6.3 as far as $Pl_0/M = 2,900$. This is useful in some circumstances: e.g. (a) tensions up to $Pl_0/M = 3,000$ may be developed naturally at higher temperatures, e.g. 20 °C, and (*b*) very high tensions are produced when a muscle is stretched during contraction, as in the experiments of Hill & Howarth (1959).

The results of Fig. 6.3 ought not to be applied, without consideration, to any other muscle than the sartorius of English *Rana temporaria*; though the evidence, so far as it goes, suggests that similar results are obtained with the sartorius of English *Bufo bufo* (1953*a*).

A problem arises of how to apply Fig. 6.3 to a muscle which is weaker than usual, or has depreciated during an experiment. Any such change must be located in the contractile component, and would not necessarily have any counterpart in the elastic one. If so, the only thing to do, except in a muscle which was obviously abnormal, or had been treated abnormally, would be to apply Fig. 6.3 in the region up to the greatest Pl_0/M of the muscle.

§3. The elastic energy

It is simple, by measuring areas in Fig. 6.3, to calculate the elastic energy present in a muscle when under any tension developed by itself. For generality the elastic energy is expressed as W/M while again the tension is expressed as Pl_0/M. The result is given in Fig. 6.4.

Consider an example: a muscle of 0·15 g with $l_0 = 3·0$ cm can exert a maximum tension of 125 g, so $P_0l_0/M = 2{,}500$. From Fig. 6.4, W/M is 23·5 g-cm/g muscle which, translated into heat units, is 0·55 mcal/g. This (1961, p. 536) is too small to explain the observed relaxation heat. Fortunately, the 'thermoelastic heat' comes to the rescue, as it did when Howarth and I (1959) had to explain the otherwise anomalous results of stretching muscles during contraction.

§4. The thermoelastic heat

When a muscle, during active contraction, is rapidly released from a higher to a lower tension, its temperature rises, apparently without any delay. The heat produced cannot be attributed to the elastic energy previously present, since that is taken out of the muscle as work by the device that lowers the tension. It was called 'thermoelastic' heat (1953*b*) because of its likeness to the effect observed, when tension is lowered, in a body with a positive coefficient of thermal expansion. In such material the release need not be rapid;

in muscle it must be, in order to avoid the after effects of release (such as redevelopment of tension) which are accompanied by chemical change.

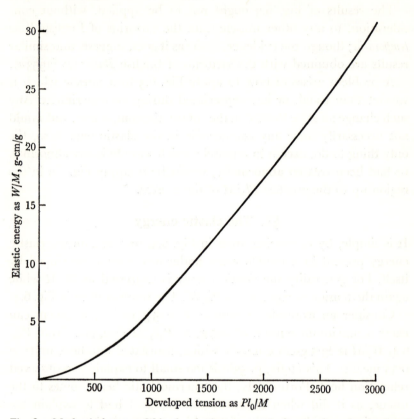

Fig. 6.4. Mechanical energy W in the elastic component of a muscle under developed tension P, calculated from the areas under the curve in Fig. 6.3. The curve is very similar to A., Fig. 2, in Hill & Howarth (1959); that was calculated from the upper curves of Fig. 8 in Jewell & Wilkie (1958).

The thermoelastic heat ΔQ following a fall of tension $-\Delta P$ was first assessed (1953 b) as $\Delta Q = 0\cdot018\, l_0(-\Delta P)$, and later by Woledge (1961) as $\Delta Q = 0\cdot014\, l_0(-\Delta P)$. With a mean 0·016, the heat produced by a fall of tension of 125 g in the assumed muscle referred to in §3 would be $\Delta Q = 0\cdot016 \times 3 \times 125 = 6\cdot0$ g-cm, which is 40 g-cm/g and nearly twice as great as the elastic energy. Together the elastic energy and the thermoelastic heat amount to 63·5 g-cm/g or about 1.5 mcal/g. No wonder that 'relaxation heat' after an isometric contraction was obvious already in 1920; particularly since the tension recording system then used was rather extensible and added substantially more energy to the genuine relaxation heat as tension fell.

No convincing evidence of the nature of the thermoelastic heat is yet apparent (though see the next paragraph). The purely thermo-dynamic view (which gave it a name), that it is caused by a fall of tension in a component of the muscle which has a rather large positive coefficient of thermal expansion, does not get one very far. Where and what is that component? It certainly must lie in the body of the muscle, the ends of which were not on the thermopile. Is it located in the contractile filaments? if so, in the thick or the thin ones? or in both where they overlap? or even in the cross-bridges? The question might be answered experimentally by finding out whether the effect is influenced by the length of the muscle. At great lengths there is less overlap, at short lengths more. A measurement, however, of thermoelastic heat must be made on a whole muscle, not a fibre, and in a whole muscle the high tension due to mechanical stretch, as distinguished from contraction, might provide too great a complication; but contractions at lengths between $1.2\ l_0$ and $0.7\ l_0$ might provide uncomplicated evidence.

R. E. Davies (1965) has speculated about the nature of this thermoelastic heat, and suggested an explanation in terms of hydrogen-bond formation. 'During a quick release of an activated muscle all bridges that were developing tension would rapidly form hydrogen bonds, contract, and give out heat. The ATP would now be split by the ATPase and the links would break with loss of tension, only to be reformed when the high relative velocity is low again.' I must leave that one to the experts.

References

Davies, R. E. (1965). *Muscle*, p. 62. A Symposium 1–4 June 1964. Pergamon Press.

Hartree, W. & Hill, A. V. (1920). *J. Physiol.* **54**, 84.

Hill, A. V. (1926). *Proc. Roy. Soc.* B, **100**, 87.

Hill, A. V. (1950). *Proc. Roy. Soc.* B, **137**, 273.

Hill, A. V. (1953*a*). *Proc. Roy. Soc.* B, **141**, 104.

6-2

Hill, A. V. (1953 *b*). *Proc. Roy. Soc.* B, **141**, 161.
Hill, A. V. (1961). *J. Physiol.* **159**, 518.
Hill, A. V. & Howarth, J. V. (1959). *Proc. Roy. Soc.* B, **151**, 169.
Jewell, B. R. & Wilkie, D. R. (1958). *J. Physiol.* **143**, 515.
Levin, A. & Wyman, J. (1927). *Proc. Roy. Soc.* B, **101**, 218.
Woledge, R. C. (1961). *J. Physiol.* **155**, 187.

7

THE TENSION–EXTENSION CURVE OF
THE SERIES ELASTIC COMPONENT (s.e.c.)
DURING ACTIVITY

§1. History and introduction

Invited by the Editor, I once wrote a letter to *Nature* (1946), in reply to one by Lord Brabazon of Tara, on 'Physiology in Horse Racing'. Recollection of the correspondence suggested a sub-title for this chapter of which I think he would have approved: 'Daylight by Obstinacy out of Disappointment'. For many years I have believed that striated muscle behaves as a two-component system, contractile and elastic in series. How otherwise could one make sense of the experiments described in ch. 1, or of a host of facts observed since? But always there have been difficulties and inconsistencies which complicated the simple picture, and writing a book forced one to face them.

Sometimes one asked, impatiently, 'does the series-elastic component really exist?' The right reply is like that to the question of whether Homer wrote the Iliad; if he did not, it was somebody else of the same name. And a name *is* necessary, in order to describe a number of well-known and reproducible facts. Let us call it the s.e.c. A more sensible question then would be, 'where is this s.e.c. located and what are its properties?' Some of it certainly resides in tendons and tendon bundles; but already in 1950, describing the earliest attempt to determine the tension–extension curve by the method of quick release from a high tension, I wrote:

but no assumption is implied [in referring to tendons] that other series elastic elements do not exist within the fibres themselves; the evidence [for the properties of the s.e.c.] is derived from mechanical experiments with active muscle, not from histological observation.

As might have been said not so long ago, the evidence for molecules is derived from chemical experiments, not from looking for them with a microscope.

A year earlier (1949, p. 415) it was conjectured that the series elasticity 'may be due in part to the elasticity of the active contractile structure' distributed along its length. If so, 'the question of its absolute length would not arise'. It seems pretty certain now that some of the s.e.c., not necessarily a constant part of it, *is* located within the sarcomeres themselves. A few experiments with single fibres, made as described in §7, might settle the matter.

In 1938 (p. 184). the first specific reference in modern terms was made to muscle as a two-component system, consisting of an elastic element in series with a contractile one. One had only vague ideas then of what the tension–length relation of the elastic element would look like; so it was natural to 'have a go' with the simplest possible model, in which extension was proportional to force. This model, together with the force–velocity relation then newly realized, allowed one (Fig. 15 there) to calculate an isometric contraction which looked rather like the real thing—apart from the fact that the muscle was assumed to be fully active from the start; one did not know any better then. The suggestion was made that, if a known compliance was put in the connector to the tension recorder, then two isometric records, one with and the other without the added compliance, would in effect provide 'simultaneous equations' by which the force–length relation of the s.e.c. could be calculated (see ch. 3, §5C). I had not realized then that from those 'simultaneous equations' the force–velocity relation also could be deduced. In 1950 Wilkie did put an extra measured compliance into the wire by which his arm muscles were connected to a tension recorder.

In 1949 (Figs. 7, 8 and 9) I tried to calculate from two experimental curves for a toad sartorius at 0 °C (a force–velocity relation and an isometric myogram) the tension–extension curve of the s.e.c. The method appeared to work quite nicely, but the result was wrong; it could not really work because the muscle was not fully active (as we know now) until 80 msec or so had passed after the start of stimulation. I had falsely concluded, from results described in the same paper, that the transition from rest to full activity is 'abrupt' (see ch. 5, §1). With a delayed isometric contraction (ch. 4, §2) it would have worked better, but that was not invented till 1967; and there are other reasons, discussed in ch. 3, §5, and in this and the following chapter, why the result would not apply generally. In 1950 Wilkie used a similar method with the flexors of his forearm; but again his muscles may not have been fully active for an appreciable interval after the start of the isometric contraction (see his Fig. 8, A(a)).

In the bad old days of the visco-elastic theory Levin & Wyman (1927) discussed a model consisting of a visco-elastic element in series with an undamped one. The visco-elastic element was really the contractile one, whose characteristic properties in 1938 became the force–velocity relation; the undamped one is the s.e.c. Curves from an experiment of theirs are reproduced in Figs. 6.1 and 6.2.

The work with Gasser, abbreviated in ch. 1, was really the origin of all my later thinking about the s.e.c. and related matters. But it seems unlikely that Adolph Fick (or his pupil Blix) was not aware of something like the s.e.c., though I have not re-read their papers to find out. Borelli even may have speculated about it 300 years ago. But obvious as the idea is, the location and quantitative properties of the s.e.c. ceased to be so obvious when one examined them in detail.

The truth is, let us face it, that the s.e.c. is not an invariable physical object (but nor is a man, though he admittedly exists); and its important characteristic, the tension–extension curve, varies according to its previous history of tension change (sooner or later, see §5 below, where the cat, or something like her, is let out of the bag). But a consistent tension–extension curve *can* be obtained by recording the tension as a function of length during rapid release of a muscle from maximum isometric tension. Details of how this tension–extension relation is determined are given in ch. 6, §2, and a mean curve for frog sartorii at 0 °C is in Fig. 6.3. The method is certainly consistent as is shown by the S.E. in that figure. The curve applies pretty well to a muscle redeveloping its tension after a quick release. It does not apply at all during the earlier stages of an isometric contraction from previous rest, before the active state is fully developed. It applies moderately well, occasionally quite well, to the later stages of any isometric contraction.

§2. Calculation of the tension–extension curve from records of isometric contraction

The method used in 1938 and 1949 (see above) was the obvious one, to use a force–velocity curve directly observed. The velocity of shortening of the contractile component (of length x) is $-dx/dt$, which can be read off from the force–velocity curve at any P; also the time t to that P can be taken from the isometric myogram. Then from $-dx/dt$ and t, by numerical integration in short enough intervals, the amount of shortening $-\Delta x$ of the contractile component up to any P can be calculated. In an isometric contraction this must be equal to Δy, the extension of the s.e.c. Δy plotted against ΔP is the extension–tension curve. No assumption is made that Δy is a function only of P; it may depend also on the manner and time in which P is reached.

Much more convenient, however, than the direct use of a force–velocity curve is to apply its equation $v = b\,(1 - P/P_0)/(1 + a/P_0)$. Even if one does not know the values of the constants b and a/P_0 it is often possible to obtain the information one wants by taking any reasonable values of them. That will be evident in what follows.

Fig. 7.1 relates to the contraction of a frog sartorius at 0 °C, which was treated in three ways.

(1) The continuous line A. is the force–length relation obtained when the muscle was rapidly released during a maximal isometric tetanus (as described in the previous chapter). It is drawn only down to 8·5 g, which was the tension at the start of redevelopment after another shorter release. This curve is from one of the eleven experiments on which Fig. 6.3 in the previous chapter was based. If Fig. 7.1 were turned round on its side counter-clockwise line A. would obviously be of a pattern with Fig. 6.3.

(2) The filled circles B. around the line were calculated from the curve of redeveloping tension after release from maximum tension during an isometric tetanus; the release ended at 8·5 g and the tension then started at once to redevelop. The muscle was already fully active, the stimulus having started 0·5 sec earlier. The constants used for calculating the distance shortened were the standard $a/P_0 = 0.25$, and $b = 0.35\ l_0/\text{sec}$; b was chosen to make the spots lie as close as possible to the line.

(3) The hollow circles C. in the upper array were calculated, *with the same constants*, from the record of an ordinary isometric contraction from previous rest. The calculation was started at a point on the record where the tension was also 8·5 g. Four of the circles are marked with numbers, which are times (msec) after the start of stimulation. The active state must have reached its full intensity somewhere between these times.

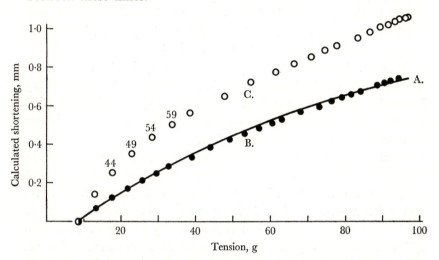

Fig. 7.1. A., the line gives the tension–extension relation of the s.e.c. obtained during rapid release from maximum developed tension; B., the filled circles give the calculated shortening during isometric redevelopment of tension after release to 8·5 g; C., the hollow circles give the calculated shortening during a normal isometric contraction from previous rest. The calculation started from 8·5 g. Numbers in C. give time in msec after the beginning of stimulation.

Conclusions can be drawn as follows:

(*a*) The shortening calculated from the record of redeveloping tension agrees pretty well with the force–length relation if a value of b is taken which is 8 % greater than the usual mean value (0·325 l_0/sec); the difference is less than the standard deviation of a single measurement of b (see ch. 3, §3).

(*b*) The shortening calculated from the record of isometric contraction from previous rest is too great to fit into any reasonable scheme. This is chiefly because at any time earlier than the moment at which the active state was fully developed the real average velocity of shortening of the fibres was less than that given by the force–

velocity relation; at short times, for example just after the latent period, the real average velocity must have been very small. There is no way of allowing for this error. It can be avoided only by using a *delayed* isometric contraction, as described in ch. 4, §2.

(*c*) The later slope of the upper array of points, after the active state has been fully developed, is rather greater throughout than that of the lower set. (This nearly always happens, but an exception is shown in Fig. 7.2 opposite.) The reason for it is that curve A. was obtained during rapid release from a high tension; and a high tension causes redistribution of length and strain in the elastic material, which is not immediately reversed when the tension is lowered. But it is reversed rapidly during the period of rest preceding an ordinary isometric contraction (see §5). In other words, a preliminary high tension makes the s.e.c. less extensible, but the greater stiffness rapidly passes off under a low tension.

Another experiment of the same kind is shown in Fig. 7.2. Release reduced the tension of the muscle to $1 \cdot 5$ g, and redevelopment started at once. The tension–extension curve is A., as in Fig. 7.1, and the array B. of filled circles was calculated from a record of tension redevelopment and a force–velocity equation. As before a/P_0 was taken as $0 \cdot 25$, but in order to get the spots near the line the value of b had to be taken as $0 \cdot 274 \, l_0/\text{sec}$. This is 16% less than the standard mean value of b ($0 \cdot 325 \, l_0/\text{sec}$); which is $1 \cdot 4$ times the S.D. of a single measurement of b. The probability of a value as low as this is $1 : 7$. As before the spots in the upper array C. were calculated with *the same constants* from the record of an isometric contraction from previous rest. Within the limits of error, C. is parallel to B. from $P = 40$ g upwards. This has happened in only one other experiment among many; and in both of them the value of b was rather unusually low.

Other experiments were made in the same way as those illustrated in Figs. 7.1 and 7.2. In one, just referred to, the best fit of the calculated distance of shortening to the tension–extension curve, during redevelopment, required a value of $b = 0 \cdot 29 \, l_0/\text{sec}$. In this experiment the later slope of the set of points C. was the same as that of the set B., as it was in Fig. 7.2. These two experiments showed a second peculiarity, which is really only another expression of the same equality as the first one. It may be of interest, however, to have it displayed, not after calculation but by a direct comparison of two records of contraction. In Fig. 7.3, the two contractions, from which the upper and lower arrays of circles in Fig. 7.2 were calculated, are shown together on the same time-base. D. is an isometric contraction

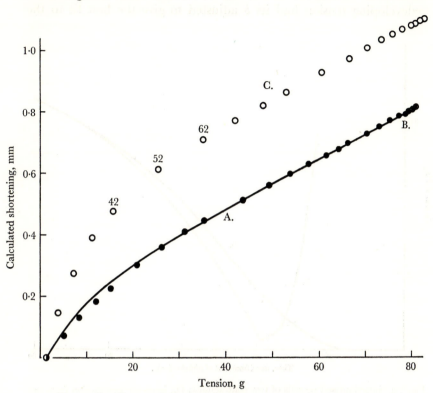

Fig. 7.2. Details are as in the legend of Fig. 7.1, except that the calculation started from a tension 1·5 g.

from previous rest, E. is a curve of redeveloping tension after a not-very-rapid release F. E.–F. was displaced in time so that it joined D. tangentially at about 87 msec. From then onwards the two coincided all the way to the summit.[1]

---

[1] At first I thought that these two experiments must be the answer to the physiologist's prayer, since muscles would be so much simpler if this result was usual. But that idea had to be abandoned. More likely, indeed, than any intervention of Providence is the fact that in both these experiments the release from high tension took rather longer than usual—which would have given more time for the visco-elastic element (§5 below) to return towards its original length. This, and possibly a shorter time constant than that used for curve B. in Fig. 7.11 below, could better explain these peculiarities.

Figs. 7.1 and 7.2 are from two of the eleven experiments from which the mean tension–length relation in Fig. 6.3 was calculated. The remaining nine were treated in the same way as illustrated in A., B. and C. of those figures, i.e. the shortening calculated from a curve of redeveloping tension had its b adjusted to give the best fit to the

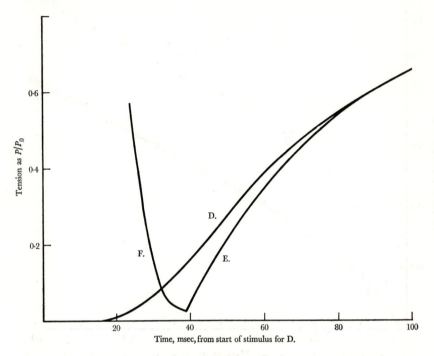

Fig. 7.3. Superimposed records of two contractions: D., isometric contraction from previous rest; E., isometric redevelopment of tension after F., a not-very-rapid release from maximum tension. The records had the same maximum and F.–E. was displaced parallel to the time base until it came into contact with D. at about 87 msec; then they ran together to the summit.

observed tension–length relation. The mean value of this b was $0.320\ l_0/\text{sec}$ with a S.D. of a single value of $\pm 0.027\ l_0/\text{sec}$. This is near the mean value (ch. 3, § 3) obtained by the usual isotonic method of determining the force–velocity relation.

Here is satisfactory evidence that the series elastic component in a muscle which has developed its full isometric tension has indeed a tension–extension relation which can be determined by a quick release in the way described in the previous chapter. The sad thing, however, is that it does not apply accurately to any contraction other than that of redeveloping tension after a quick, but not too quick, release.

It would have been better if I could have made actual force–velocity curves by the isotonic method to compare with the results described above. But these experiments were already long ones, and the change round of apparatus necessary in order to make such records might have given time for the muscle to depreciate. Fatigue of the muscle (and of the observer!) could have deprived the comparison of any significance. Anyhow it was not done: the present post-mortem on experiments made three years ago is all that can be offered.

The reader will have noticed that throughout the argument I have assumed a uniform value of $a/P_0 = 0.25$. In fact, of course, a/P_0 varies from one frog sartorius to another and it is shown in ch. 3, § 3, that the S.D. of a single value is about 12.5% of the mean. But the effect of a small alteration of a/P_0 on an analysis, like that which gave Figs. 7.1 and 7.2, cannot easily be distinguished from that of a small change of b. This is shown by the three curves in Fig. 7.4. These represent the analysis of a record of tension redevelopment with a common value of b (10 mm/sec) and assumed values of a/P_0: for A., 0.208; for B., 0.25; for C., 0.30. The value for B. is 20% greater than for A.; for C. it is 20% greater than for B. Now 20% is 1.6 times as great as the S.D. mentioned above, which is a fairly substantial amount. Nevertheless, by multiplying the ordinates of A. by 0.92, and those of C. by 1.09, the three curves can be made practically indistinguishable. Expressed in a general way, the effect of a given percentage change in a/P_0 can be compensated by about half that percentage change in b.

This, of course, is only an approximation; but its consequence is that for the kind of analysis shown in Figs. 7.1 and 7.2, and elsewhere, one would generally be wasting time in trying to distinguish between the effects of the two constants. This does not mean that the force–velocity equation can be expressed in terms of one constant only—it certainly requires two. But it does mean that, in studies of the form of different kinds of contraction under varying tension, one can usually start with a standard value of a/P_0 and proceed as though only b was adjustable. The error caused by so doing is more significant during the early parts of a contraction during which tension is low. This is easily seen from the equation $v = b(P_0-P)/(P+a)$; when P is small an error in a has a larger effect on v.

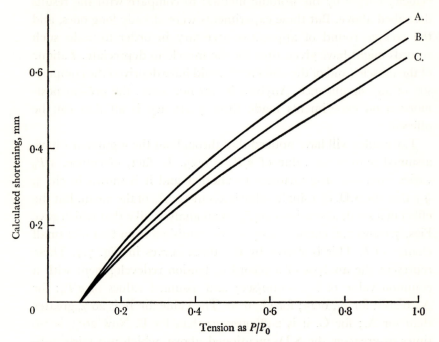

Fig. 7.4. Shortening during isometric redevelopment of tension after release, all calculated with the same value of b, but with three different values of a/P_0. See text.

Frequent reference has been made to the delayed isometric contraction. An example of this was shown in Fig. 4.3, where a detailed description was given in the text. Its results have been analysed in the same way as was done for Figs. 7.1 and 7.2, the calculated shortening (Fig. 7.5) of the contractile component being plotted against the tension developed, B. for the delayed isometric contraction, C. for an isometric contraction from previous rest. The same difference is evident in Fig. 7.5, as in Figs. 7.1 and 7.2 between the arrays B. and C. In C. the active state took time to develop, in B. it was fully developed at the start. But the calculated shortening in B. Fig. 7.5

94

was greater than in its nearest equivalent B. Fig. 7.1, because for the latter the s.e.c. had been stiffened by a previous high tension; which does not happen in a delayed isometric contraction. So B. Fig. 7.5 is the nearest approach to the s.e.c. of the muscle in this sort of contraction.

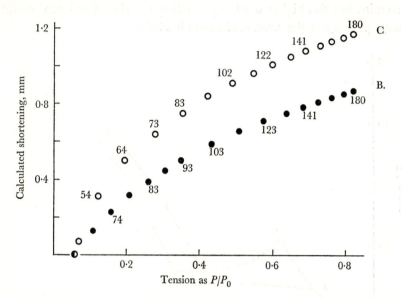

Fig. 7.5. Calculated shortening, B. during a delayed isometric contraction and C. during an isometric contraction from previous rest. The numbers near some of the circles are times in milliseconds from the start of stimulation.

§3. Stiffening of the s.e.c. by a developed tension

The stiffening of the s.e.c. by a developed (not an applied) tension is illustrated in Fig. 7.6. A muscle was stimulated isometrically at length $l_0 + 0.5$ mm and released 1.5 mm at 24.6 mm/sec (0.86 l_0/sec). The speed was low enough for practically all the fibres to keep up. In one contraction A. release was at 33 msec after the start of stimulation and 'stop' at 94 msec. In the other contraction B., made 10 min later, release was at 52 msec and 'stop' at 113 msec. The maximum tensions were the same. The characteristic to be noted is that, at any tension, the rate of rise in B., after 'stop', is about 15 % greater than in A., and continues greater throughout as shown by the convergence of the curves. The active state must have been fully developed a long time before either 'stop' occurred. The difference of slope is not large, but has been checked on the original records. The slope dP/dt at any P depends on the constant b of the force–velocity relation and the compliance dy/dP of the s.e.c. One can see no reason why the force–velocity relation should have changed within 10 min, but the higher tension preceding the rise after 'stop' could have pulled out the s.e.c. and made it stiffer.[1]

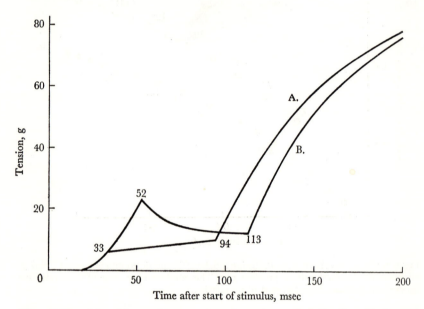

Fig. 7.6. Two delayed isometric contractions, A. made 10 min before B.; with the same interval of release but starting and ending 19 msec later in B. than in A. The slope dP/dt, at any tension, is greater throughout in B. Maximum tension 101 g. See text.

This effect of a higher tension on the s.e.c. was small, but the extra tension was small. A greater effect has continually been observed when curves of redeveloping tension after release from a higher tension were compared with ordinary isometric curves after previous rest. At times after the start of stimulation, when the active state must have been fully developed in the isometric contraction, the slope dP/dt of its curve at any P was less than it was in the curve

---

[1] It is interesting that in Fig. 7.6 the two curves of release seem to be converging, from above and below, to a common tension of 11 g. The maximum tension was 101 g, so, from the force–velocity equation, $24 \cdot 6 = 90b/(11 + 25 \cdot 25)$; this gives $b = 9 \cdot 9$ mm/sec $= 0 \cdot 35\ l_0$/sec which is near the usual value.

of redeveloping tension at the same P. In a fully active muscle $dP/dt = v\,dP/dy$, where v is the velocity of shortening, assumed to depend only on P. So if dP/dt is greater, dP/dy must be greater, i.e. the s.e.c. is stiffer. The stiffness of the s.e.c. is not simply a function of P, it depends also on the immediate previous history of tension in the muscle. This is a misfortune for the physiologist, it would have been easier for him if dP/dy were a function only of P; but few effects in physiology are as simple as that.

§4. The linearity of the tension–extension curve in the higher range of tensions

It has been shown above that the tension–extension curve of the s.e.c. can be calculated from the record of an isometric contraction if the force–velocity relation is known, or assumed. A curious regularity has been found in the tension–extension curve of a frog sartorius so obtained; *it becomes rather accurately linear at tensions above about* $0\cdot5\,P_0$. In a toad sartorius the range of linearity is greater, from $0\cdot35\,P_0$ or $0\cdot4\,P_0$ upwards; indeed it begins at about the time when, on other evidence, the active state has reached its full intensity. The same linearity is found with tortoise muscles.

Many records have been made, with frog sartorii at $0\,°C$, of the tension in isometric contractions, either from previous rest or after release from a high tension. These were analysed in the way described in §2, to give the shortening $-\Delta x$ of the contractile component (which is the same as the stretch Δy of the s.e.c.) as a function of P the tension developed. The slope of the lines of Δy against P is referred to as $\Delta y/\Delta P$, and is given in mm/g tension. The constants of the force–velocity equation were taken (in the case of frog sartorii) as $b = 0\cdot325\,l_0/\mathrm{sec}$ and $a/P_0 = 0\cdot25$. *But the linearity does not depend on the values chosen.*

In Fig. 7.7 are six lines obtained by calculation from records of tension redevelopment after release from maximum developed tension. Tension is given vertically on a common scale (though see p. 98 for line *a.* on the left); shortening horizontally on a scale marked above in units of $0\cdot1$ mm, but with no common origin. Maximum tension is shown by a bar above the end of each line, half maximum tension by a bar crossing the line lower down.

The spots, representing the calculated values of Δy, start (on the left) above the lines later reached, but come on to them at about $0.5\ P_0$.[1] The peculiarities at the upper ends may be due simply to the difficulty of reading the records accurately near their maxima; but the upward deviations found in three of the curves could have been caused by redistribution of length between sarcomeres under a high tension (see ch. 8). Anyhow the deviations at the top are small (not more than 20 μ) and not consistent.

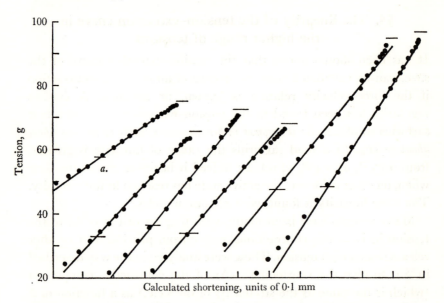

Fig. 7.7. Frog sartorii at 0 °C in isometric redevelopment of tension after release. Calculated shortening of the contractile component horizontally, in units of 0·1 mm (scale below), against tension vertically, scale on left. The differences of slope are largely due to differences of muscle size; *a.* on the left is for a very small muscle.

The calculations were made from copies of the records but, for accuracy, allowance is required for the compliance of the connecting wire. A more important allowance, however, is necessary for the dimensions of the muscles. Line *a.* on the left (zero tension for this is at the point marked 40) was obtained with a very small muscle ($P_0 = 35$ g), and the small slope $\Delta y/\Delta P$ is due to this. The dimensions can be allowed for as follows:

(1) In muscles of different lengths, other things being equal,

[1] The axes of P and Δy are the reverse of those in the figures of §2.

Δy must be proportional to l_0; but that is already taken care of by using for each calculation a value of b which is proportional to l_0;

(2) in muscles of different cross-sections ΔP must, in general, be proportional to the cross-section, which depends on M/l_0.

Thus $\Delta y/\Delta P$ should vary, everything else being equal, as l_0/M. To take account of this, we divide by the actual l_0/M and multiply by the l_0/M of some 'standard' muscle. The standard muscle adopted had constants which were the average of those for the nine muscles referred to next: $M = 106$ mg, $l_0 = 29$ mm, $l_0/M = 0.274$, $P_0 = 73$ g. In these nine experiments the mean 'standard' value of $\Delta y/\Delta P$, during redeveloping tension after rapid release, was 0.0075 mm/g with a S.E. of the mean of $\pm 6.4\%$. The S.E. includes that due to possible errors in the assumed values of b.

The same linearity as is pictured in Fig. 7.7 is found also in ordinary isometric contractions after previous rest. In fourteen experiments made between February and August the mean 'standard' value of $\Delta y/\Delta P$ calculated in the same way between the same limits (0.5 P_0 to P_0) was 0.0089 mm/g ($\pm 5.2\%$, S.E. of mean).

The ratio of these two means is 1.19. In eight of the experiments the ratio was measured with *the same muscle;* its mean was 1.16. Thus, on the average, the compliance of the s.e.c. between 0.5 P_0 and P_0 was about 17% greater in an ordinary isometric contraction than in an isometric redevelopment of tension after release. *This is convincing evidence of the stiffening of the s.e.c. by an immediately preceding exposure to a high developed tension.*

A. *Toad sartorii at* 0 °C

In 1965, a year earlier than the experiments described above, many records were made of twitches of toad sartorii, for the purpose of finding out how long an interval is required, after a shock, for the active state to be fully developed. These experiments are described in ch. 5, §3. In order to check the earlier finding (1949) that the tension can be raised by an appropriate stretch during a twitch to that in a maximal tetanus, a number of records of isometric tetanic contractions were made. One of these[1] is shown in Fig. 7.8 with

---

[1] This record shows the effect described in ch. 9, §2, that after the stimulus was over the tension fell normally to about 0.1 P_0, and then very slowly to its initial level.

3 sec sweeps: the spots where visible are at intervals of 20 msec. After I had finished, in 1968, the analysis reported above of the records made in 1966 with frog sartorii, it occurred to me that the earlier records with toad muscles might provide an interesting comparison with the results for frogs. Many of the toad records had been made with quicker sweeps, and with spots (every 20 msec) all distinguishable; so the times were very accurate.

The results, all at 0 °C, were interesting; not only was the relation between Δy and ΔP rather exactly linear, but the linearity continued

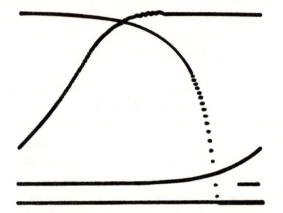

Fig. 7.8. Isometric tetanic contraction of toad sartorii at 0 °C. Note the peculiar relaxation. Read from right to left. Original base line at bottom.

down to lower tensions than in isometric contractions of frog muscles. Indeed, in the six experiments analysed, the time at which the relation between Δy and P became linear averaged about 90 msec after the start of stimulation; in ch. 5, § 3, it was shown that, in toad sartorii at 0 °C after a single shock, the active state is fully developed at about 80 msec.

The results of four experiments are shown in Fig. 7·9. The abscissa is P/P_0, the ordinate is Δy. The value of b in the force–velocity equation, assumed for the calculation, was 6 mm/sec, which, taking account of muscle length, is about one half the value assumed for the frog calculations described earlier; this is roughly in proportion to their intrinsic speeds. The linearity is not affected by the choice of b.

The lines in Fig. 7.9 have not been corrected for the compliance of the connector joining the muscle to the tension recorder. When that was done the average value of $\Delta y/\Delta P$ in the six experiments was 0·0066 mm/g. But that cannot be compared directly with the average value for frog muscles reported above; the toad muscles were longer and thicker. So the dimensional method described above was used

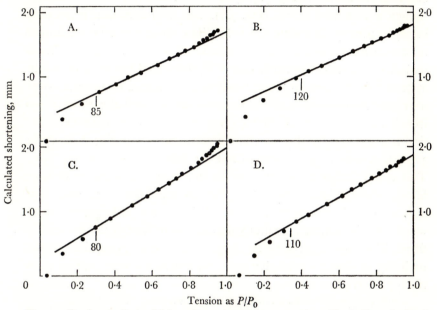

Fig. 7.9. Toad sartorii at 0 °C in ordinary isometric contractions. Vertically, calculated shortening of the contractile component; horizontally, tension as P/P_0. The numbers are times after the start of stimulation (msec) at which the relation became linear.

to bring the results to a mean 'standard' value, as was done with the frog muscles. The answer then, for the same 'standard' muscle, is $\Delta y/\Delta P = 0\cdot0115$ mm/g, for the average of the six experiments in fourteen frog experiments it was 0·0089. It is satisfactory that the same linear relation between Δy and ΔP is found in toads as in frogs; and interesting that the linearity starts relatively earlier in the toads. Perhaps that may be related to what I have a suspicion may be a fact, that toad muscles have not so wide a dispersion of their individual fibre speeds as frog muscles.

In the six toad muscles the average maximum value of dP/dt early in contraction was $5 \cdot 5$ P_0/sec. In frog muscles it is usually, in normal isometric contractions, about 10 P_0/sec. This is indirect evidence that the ratio of the b's is about $0 \cdot 5$.

The toad sartorii were very strong, having an average value of $P_0 l_0/M$ of 2380. Perhaps for that reason the high-tension end of the $\Delta y : \Delta P$ line is apt to turn up, in some experiments more than in others as illustrated in Fig. 7.9. This occurs sometimes in frog muscles though to a smaller extent (Fig. 7.7). The final deviation is not very large, its average is less than $0 \cdot 1$ mm. The effect could be attributed to the gradual extension of a minority of weaker sarcomeres by a majority of stronger ones. If this really occurred the tension would approach its maximum more slowly and Δy would be summed over a greater number of equal time intervals to reach the same P; so its calculated value would be too great. But another possible explanation is suggested by the peculiar later relaxation in Fig. 7.8. Early relaxation here is quite normal but the last 10 % of it is very slow. This could indicate the presence in the muscle of a group of slow fibres, which might continue to shorten after the faster ones had developed their maximum tension. Their shortening being very slow would require a much smaller value of b than that of the rest of the fibres. To calculate it with the larger value of b appropriate to the majority of fibres would give an apparent shortening greater than really occurred.

B. *Rectus femoris of tortoise*

In Table I of Woledge's paper (1968) are the results of eight experiments on isometric tetanic contractions of tortoise muscles at 0 °C. In five of these Dr Woledge was able to calculate the shortening of the contractile component in the way described above for frogs and toads; and he has allowed me to use his results. These are:

(1) The relation between $-\Delta x$ and P was strikingly linear over the range (on the average) from $0 \cdot 33$ P_0 to near P_0.

(2) The average value of $-\Delta x/\Delta P$, after allowance for length and weight of muscle as described above, was about $3 \cdot 8$ times as great as for frog muscles or $3 \cdot 0$ times as great as for toad muscles.

For the calculation, the values of b and a/P_0 were those determined in each experiment. The mean value of b was $0 \cdot 0154$ l_0/sec, of a/P_0 it was $0 \cdot 07$.

§5. A visco-elastic element in the s.e.c.

It was shown in §4 that within a considerable range of developed tension (frog $0.5\,P_0$ upwards, toad and tortoise $0.35\,P_0$ upwards) the compliance of the s.e.c. *behaves as though* it were constant. The limiting phrase is necessary because the compliance appears to depend, to a significant extent, on its previous history of stress and strain. It was shown, for example, that the apparent compliance (in the higher tension range) is, on the average, about 17 % greater during an isometric contraction from previous rest than during isometric redevelopment of tension after rapid release from maximum tension. And there is other evidence (§§2 and 3 above) of stiffening of the s.e.c. under the influence of a developing tension. All these effects suggest *that developing (or developed) tension causes visco-elastic stretch in the s.e.c.* This would not require a separate element there, the visco-elasticity could be distributed throughout the main elastic component (whatever that is). It is simpler, however, to consider it separately as follows.

A muscle is stimulated isometrically, and during equal short intervals c, beginning at the moment when the tension starts to rise, increments of tension are developed as follows:

$$p_1, p_2, p_3, \ldots, p_n, \ldots$$

The total tension therefore at time nc is the sum of these. For purposes of calculation each increment p can be regarded as located at the middle of its interval. Any one of them, say p_r, is supposed to have two effects:

(1) an immediate elastic stretch of the s.e.c., and

(2) a gradual visco-elastic stretch proportional, at time t after the middle of its interval, to $p_r\,(1 - e^{-t/t_0})$; for simplicity call this $p_r F\{t\}$; the constant of proportion can be introduced later.

These effects are supposed to be independent of one another, and of the consequences of the other increments p.

At time nc the visco-elastic stretch resulting from p_1 is $p_1 F\,\{(n - \frac{1}{2})c\}$, that resulting from p_2 is $p_2 F\{(n - \frac{3}{2})c\}$, from p_3 it is $p_3 F\{(n - \frac{5}{2})c\}$ ... from p_{n-1} it is $p_{n-1}F\{3c/2\}$, from p_n it is $p_n F\{c/2\}$. The total stretch at time nc is, writing it out in full:

$$p_1 F\{(n - \tfrac{1}{2})c\} + p_2 F\{(n - \tfrac{3}{2})c\} + \ldots + p_{n-1}F\{3c/2\} + p_n F\{c/2\}. \quad (1)$$

The calculation is similar to that described in my book (1965) pp. 311–15, for the 'multiplication of series'.

An ordinary isometric contraction (P_0 = 100 g) was used to provide the data, p_1, p_2, p_3,...being read off at suitable small intervals. The results are given in curves A. and B. of Figs. 7.10 and

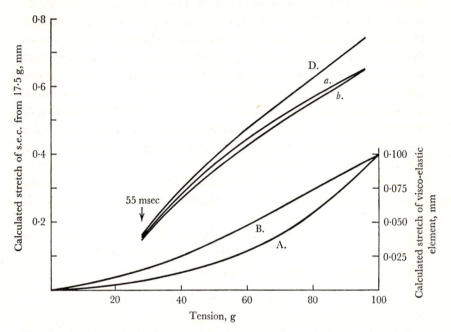

Fig. 7.10. A. and B., calculated stretch of the visco-elastic element in the s.e.c., from P = 0, during an isometric contraction; A. with time-constant 50 msec, B. with 16·7 msec. D. calculated stretch of the entire s.e.c. from 17·5 g upwards (as for Figs. 7.1 and 7.2); a. and b. the same, but after allowing for the stretch of the visco-elastic element.

7.11. For A. the time constant t_0 was taken as 50 msec, for B. 16.7 msec. For both the total visco-elastic stretch, from P = 0 to P = P_0, was taken as 0·1 mm. For A. the stretch due to any increment of tension p would be 99 % complete in about 230 msec, for B. in about 77 msec.

Curve D. in Fig. 7.10 is the stretch of the whole s.e.c. during the same isometric contraction, calculated as in §4 and plotted against P. The calculation began at 17·5 g (45 msec). It shows the usual region of linearity, but the result is mildly exceptional in the fact that linearity began at about $P = 0.6 P_0$ instead of $0.5 P_0$. Line D. gives a measure of the total stretch of the s.e.c. during an isometric contraction from 17.5 g upwards, and that must include the visco-elastic stretch shown in A. or B. (note that these are plotted on four times the scale of D.). If the constants are chosen correctly the purely elastic element in D. can be obtained by subtracting the ordinates of A. (or B.) from the ordinates of D. That gives line a. (or b.) which, on these assumptions, represents the purely elastic element in the s.e.c.

All this may seem rather fanciful, particularly as only an approximate estimate can be made of the constants. But it gave me much consolation after struggling for a long time to understand what was happening. In Fig. 7.10 line D., at any time after the active state has been fully developed, represents the tension–extension curve of the effective s.e.c. during the contraction concerned; while a. (or b.) is the tension–extension curve applicable to a contraction in which the visco-elastic element has effectively been eliminated during a previous stretch by maximum developed tension. That is, a. (or b.) should be the tension–extension curve applicable to isometric re-development after rapid release from a high to a low, but not to zero, tension. Lines a. and b. differ from D. by about the amount found experimentally.

That led to a further inquiry. How rapidly, in the suggested model, would the strained condition caused by maximal developed tension disappear during release? The same method of calculation, and data from the same experiment, were used as gave the curves of Fig. 7.10. A muscle was stimulated isometrically and when maximum tension had been developed it was rapidly released and the tension fell in 10 msec to zero (C., Fig. 7.11). A calculation similar to that described above was applied to the falling phase of tension but this time in intervals of 1 msec. For A. the time-constant was 50 msec as before, for B. it was 16.7 msec; and the total visco-elastic shortening between 100 g and 0 g was chosen as 0.1 mm.

Curve C. is one of the eleven of which Fig. 6.3 in ch. 6, § 3, gives the mean. During the 7.4 msec required for the tension to fall to 5 g the calculated shortening of the visco-elastic element was 0.007 mm (7 %) in curve A., 0.021 mm (2 %) in curve B. In 230 msec, however, with the longer time-constant, the visco-elastic element lengthened by its full amount, 0·1 mm; this took only 77 msec with the shorter time constant. Thus in an ordinary isometric contraction from previous rest the s.e.c. would include 0·1 mm of the visco-elastic element starting fully relaxed; while in redevelopment of tension after quick release to 5 g the s.e.c. would include only 7% of it

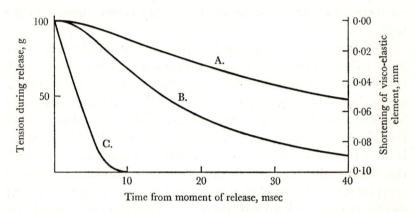

Fig. 7.11. C., rapid release of tension during maximal isometric contraction (scale on left). A. and B., calculated shortening of the visco-elastic element during and after release; A., with time constant 50 msec; B., with 16·7 msec.

($t_0 = 50$ msec) or 21 % of it ($t_0 = 16\cdot7$ msec). This gives a credible explanation of the difference found in §4 between ordinary isometric development of tension from previous rest and isometric redevelopment of tension after release.

The same sort of calculation can be applied to other kinds of contraction.

§6. The logarithmic form of the isometric myogram in the higher tension range

The linear relation between Δy (or $-\Delta x$) and P, described in §4, accepted merely as an empirical fact, has an interesting consequence: the time t is a simple logarithmic function of the tension P. If h be the slope of the line of $-\Delta x$ against ΔP, we may write:

$$h = -dx/dP = -(dx/dt)/(dP/dt) = v/(dP/dt).$$

If the usual force–velocity equation is used this becomes:

$$(b/h)\,dt/dP = (P+a)/(P_0-P).$$

The integral of this is

$$bt/hP_0 = -(1+a/P_0)\ln(1-P/P_0) - P/P_0 + B,$$

where B is a constant. With common logarithms, and $a/P_0 = 0.25$, this becomes:

$$bt/hP_0 = -2.88\log(1-P/P_0) - P/P_0 + B. \qquad (1)$$

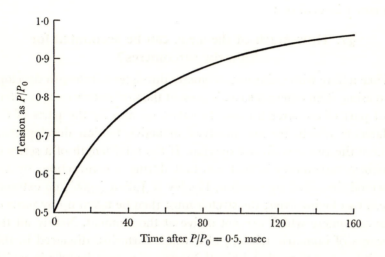

Fig. 7.12. The form of the isometric myogram of a frog sartorius at 0 °C, in the range of tension from 0.5 P_0 upwards: calculated from eqn 1 in the text.

If the zero of t is taken at the moment when $P/P_0 = 0.5$ (which is where linearity usually begins in frog sartorii at 0 °C) the constant B is -0.366. If $t = 0$, where $P/P_0 = 0.35$ (which is where linearity generally starts with toad sartorii at 0 °C), B is -0.189. With tortoise muscle, beginning at $P/P_0 = 0.35$, but with a mean value of $a/P_0 = 0.072$ (Woledge, 1968), B is -0.111.

For a 'mean' frog sartorius ($l_0 = 29$ mm, $M = 106$ mg, $P_0 = 73$ g) at 0 °C it was shown in §4 that $h = 7.5 \times 10^{-3}$ mm/g. If b is taken as $0.325\,l_0$/sec $= 9.42$ mm/sec, hP_0/b becomes 0.058 sec. With this value of hP_0/b the curve of Fig. 7.12 has been calculated from $P = 0.5\,P_0$ upwards. For any other muscle, take its values of

P_0 and b, and calculate h as $(7\cdot5 \times 10^{-3}\,\mathrm{mm/g})\,(l_0/29)\,(106/M)$. An explanation of this is given in §4, p. 99. Somebody might find in this an amusing 'numerical experiment' to make. The best way to make it is to take an ordinary respectable isometric myogram and calculate, for various values of $P/P_0 > 0\cdot5$, the right-hand side of eqn 1. This should be a linear function of t; the slope of the line is hP_0/b. Of course, allowance should be made for the compliance of the connector before the calculation is made. If the compliance is a linear function of P, say p mm/g, the easiest thing to do is to write $(h+p)$ in eqn 1 instead of h.

§7. How much of the s.e.c. can be accounted for within the sarcomeres?

When a fibre with uniform sarcomeres throughout its length develops a tension, if an external s.e.c. is present the sarcomeres must shorten. But part of the stretch could be taken up also by the parts of the filaments which are not involved in active tension development, where the two sets do not overlap. If the total length of a series of uniform sarcomeres is kept constant during a contraction, by the method described by Gordon, Huxley & Julian (1966), no external s.e.c. can be involved; the stretch must then be taken up by parts of the sarcomere which are not active at the moment. Nearly all the records of isometric tension, of a whole sartorius, discussed in this book have been made within the range of muscle lengths in which the maximum developed tension has its greatest value. This range for fibres of the frog semitendinosus, according to Gordon, Huxley & Julian (1966), centres at a sarcomere length of about $2\cdot15\,\mu$; the diagram of their Fig. 14 shows 20–25 % of the length of a $2\cdot15\,\mu$ sarcomere, say $0\cdot5\,\mu$, not engaged in active tension development.

During a delayed isometric contraction of a frog sartorius at $0\,^\circ\mathrm{C}$, the calculated shortening (§2 above) of the contractile component, between low and maximum tension (assuming an inextensible connector), is about $0\cdot025\,l_0$. This corresponds to $0\cdot054\,\mu$ in a $2\cdot15\,\mu$ sarcomere, which is 11 % of the $0\cdot5\,\mu$ available for stretch (see above). In Figs. 1, 4 and 6 of Gordon, Huxley & Julian's paper are records of delayed isometric contractions of a single fibre, with no external s.e.c., at 3–6 °C. They look very like delayed isometric contractions of a frog sartorius, but their scale of time is too small to make an accurate comparison. If they were available on a larger scale, they could be compared with records of delayed isometric contractions of a whole muscle, made under comparable conditions at the same temperature. Then we could answer the question at the head of this section.

References

Gordon, A. M., Huxley, A. F. & Julian, F. J. (1966). *J. Physiol.* **184**, 170.

Hill, A. V. (1938). *Proc. Roy. Soc.* B, **117**, 136.

Hill, A. V. (1946). *Nature,* **185**, 674.

Hill, A. V. (1949). *Proc. Roy. Soc.* B, **136**, 399.

Hill, A. V. (1950). *Proc. Roy. Soc.* B, **137**, 273.

Hill, A. V. (1965). *Trails and Trials in Physiology.* London: Edward Arnold.

Levin, A. & Wyman. J. (1927). *Proc. Roy. Soc.* B, **101**, 218.

Wilkie, D. R. (1950). *J. Physiol.* **110**, 249.

Woledge, R. C. (1968). *J. Physiol.* **197**, 685.

8

INTERNAL REDISTRIBUTION OF LENGTH IN A MUSCLE DURING ISOMETRIC CONTRACTION

§1. Introduction

It is almost inconceivable that a muscle fibre containing 10,000 more-or-less-independent contractile units (the sarcomeres), in series with one another, should be equally strong throughout its length; more explicitly, that the 10,000 sarcomeres should be able independently to develop the same maximum tension during stimulation. The fibre is taken to be 22·5 mm long, each sarcomere being 2·25 μ. If the sarcomeres are not equally strong the stronger will stretch the weaker. There is in fact a built-in stabilizing mechanism tending to counter this effect; but it is not absolute, it reduces though does not abolish it (ch. 9, §3). When a muscle, during maximal contraction and exerting tension P_0, has its tension *reduced* by a small amount, ΔP, it begins to shorten at once with velocity v (calculable from the force–velocity relation); but if its tension is *increased* by ΔP, it begins to lengthen with a velocity considerably less than v. According to Katz (1939), in a frog sartorius at 0 °C the rate of lengthening in the second case is about one fifth of the rate of shortening in the first case.

Suppose that in a fibre of length l_0 half the sarcomeres can exert a maximum force P_1, the other half a smaller maximum force P_2. The total length of the P_1 sarcomeres is $l_0/2$, so is that of the P_2 sarcomeres. We use the force–velocity equation, assume that a/P_0 is the same in all the sarcomeres ($P_0 = P_1$ or P_2), that with $P < P_2$ b is $kl_0/2$ for both groups of sarcomeres, for $P > P_2$ b is $kl_0/2$ for the stronger group and 0·2 $kl_0/2$ for the weaker group. For calculation, take the common a/P_0 as 0·25.

§2. What is the steady force exerted by a non-uniform fibre during an isometric contraction?[1]

This is clearly less than P_1 and greater than P_2. Since the contraction is isometric the velocity of shortening of the stronger sarcomeres is equal to the velocity of lengthening of the weaker ones; so, from the force–velocity equation:

$$(kl_0/2)(1 - P/P_1)/(P/P_1 + 0\cdot25) = (0\cdot2\,kl_0/2)(P/P_2 - 1)/(P/P_2 + 0\cdot25).$$

If P_2/P_1 is taken as $0\cdot8$, this equation reduces to

$$1\cdot2\ P^2/P_1^2 - 0\cdot91\ P/P_1 - 0\cdot24 = 0.$$

The solution of this is $P/P_1 = 0\cdot965$; so the force exerted by the stronger sarcomeres is reduced by $3\cdot5\%$ while stretching the weaker ones. But if $P = 0\cdot965\ P_1$ it is also $1\cdot21\ P_2$; so the tension in the weaker sarcomeres is increased 21% during stretch by the stronger ones. The velocity of shortening of the stronger sarcomeres, calculated from the force–velocity relation, is $(kl_0/2)(P_1 - 0\cdot965\ P_1)/(0\cdot965\ P_1 + 0\cdot25\ P_1)$; this is $0\cdot0288\ kl_0/2$, which must also be the velocity of lengthening of the weaker sarcomeres.

This, of course, is a very simple case. In nature the P_0's of the individual sarcomeres might follow a probability curve of some kind; but the labour of working out the consequence of that would be ill spent, since the general manner in which the system works is easily understood from simple examples.

§3. How is the force–velocity curve changed by half the sarcomeres being weaker than the rest?

Three cases are considered: (1) a uniform fibre with $P_2/P_1 = 1$; (2) the same as in §2 above with $P_2/P_1 = 0\cdot8$; (3) an exaggerated example with $P_2/P_1 = 0\cdot5$. Case (3) is not likely to occur in real life, except in an animal in poor condition; it is discussed here because

---

[1] b is the velocity constant of the force–velocity equation and k is a number which, in a whole sartorius at 0 °C, is taken as $0\cdot325\ l_0$/sec. The condition is never really steady because the stronger sarcomeres would go on stretching the weaker ones for a long time until the effect of length change intervened; but the process is slow when the two groups of sarcomeres are nearly of equal strength and for a time the tension would be nearly constant.

the difference between a uniform fibre and one with $P_2/P_1 = 0.8$ cannot easily be recognized visually in a drawing of usual size.

The calculation is similar to that in §2 and the results are given in Fig. 8.1. Curve 1. is a force–velocity curve for a muscle with uniform sarcomeres throughout $(P_2/P_1 = 1)$, curve 2. is for case 2.

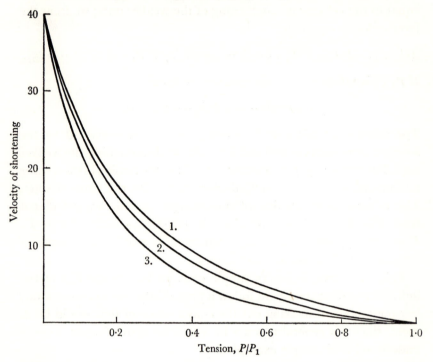

Fig. 8.1. Calculated force–velocity relation: 1., for a muscle fibre with similar sarcomeres throughout; 2., for a muscle fibre in which half the sarcomeres have maximum tension P_1, the other half maximum tension $0.8\ P_1$; 3., for a muscle fibre in which half the sarcomeres have maximum tension P_1, the other half maximum tension $0.5\ P_1$.

$(P_2/P_1 = 0.8)$, curve 3. for case 3. $(P_2/P_1 = 0.5)$. Points to be noted are:

(a) In curve 3. there is a visible discontinuity of slope at $P = 0.5\ P_1$, above this the weaker sarcomeres are stretched; in curve 2. there is a similar but less visible discontinuity at $P = 0.8\ P_2$.

(b) Curve 1. reaches the horizontal axis at $P = P_1$, curve 2. reaches it (as in §1) at $P = 0.965\ P_1$, curve 3. reaches it at $P = 0.908\ P_1$; to the right of the second and third points the velocity becomes negative, but not enough to be shown properly on the small scale of Fig. 8.1.

(c) At $P = 0$ the velocities converge to the same value; this is simply because we have assumed both sets of sarcomeres to have the same a/P_0.

§4. How is the shape of the isometric myogram affected by inequality between two groups of sarcomeres?

We compare case 1. $(P_2/P_1 = 1)$ with case 2. $(P_2/P_1 = 0.8)$. It has been shown already in §2 that the tension maxima differ by 3.5 %; but more important differences than that occur. For the calculation we assume that the fibre is already fully active and use an actual (but scaled down) tension–extension curve of a whole muscle, as obtained by quick release from maximum developed tension (Fig. 6.3). During an isometric contraction the relation holds:

$$dt/dP = (dy/dP)/v. \tag{1}$$

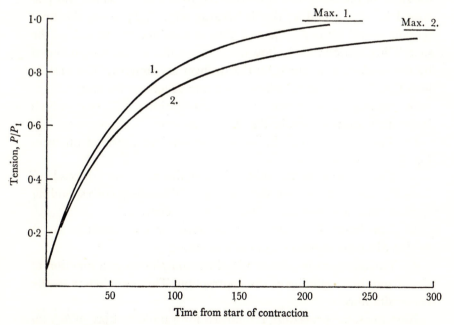

Fig. 8.2. Calculated curves of isometric tension development, for the two fibres whose force–velocity curves are 1. and 2. in Fig. 8.1. The maxima of the two curves are shown above their ends.

Here v is the velocity of shortening of the contractile component and dy/dP is the slope of the tension–extension curve, both at tension P. Now v can be read off from Fig. 8.1, curve 1. or curve 2., at any P. If this is substituted in eqn 1, dt/dP is obtained as a function of P; from which, by numerical integration, the relation between P and t can be calculated. In Fig. 8.2. the upper curve 1. is the isometric myogram so obtained for a uniform fibre, $P_2/P_1 = 1$; while the lower curve 2. is for a fibre in which $P_2/P_1 = 0.8$. Curve 2. converges slowly to a maximum $0.965\,P_1$ as described in §1, curve 1. not so slowly to a maximum of P_1.

§5. How is exchange of length shared between weaker and stronger sarcomeres during an isometric contraction?

For the calculation of §4 a series elastic component was assumed which would allow a shortening of about 0.8 mm in a normal fibre with $l_0 = 31$ mm. The total amount of shortening would be nearly the same in the composite fibre referred to as case 2., which was discussed in §§2, 3 and 4; this is because with $P_2/P_1 = 0.8$ the final tension would be $0.965\,P_1$, and that determines the stretch of the s.e.c. The question then is: how much of this shortening is contributed by the weaker, how much by the stronger sarcomeres? The answer is in Fig. 8.3 and the calculation is made as follows for case 2.

(a) A table is constructed for tensions (multiples of P_1) 0.05 (which is the start of the curves in Fig. 8.2), 0.10, $0.15, \ldots, 0.90$, 0.93.

(b) The time intervals between these tensions are read off from Fig. 8.2, curve 2.

(c) The tensions at the mid-points of the intervals are read off from Fig. 8.2, curve 2., and the velocities of shortening corresponding to those tensions are read off from Fig. 8.1, curve 2.

(d) The intervals in (b) are multiplied by the velocities in (c) to give distances shortened in each interval.

(e) The distances in (d) are added up and plotted as a function of time from the beginning of the contraction. The result refers to the composite fibre.

Next the same procedure is followed for a uniform fibre of *half the length* (maximum tension P_1), with the only difference that in (c) the velocities are read off from curve 1., not curve 2., of Fig. 8.1. The result refers to the stronger sarcomeres of the composite fibre. For the weaker sarcomeres the result is then obtained by subtraction.

In Fig. 8.3, curve *a.* refers to the stronger sarcomeres alone, curve *b.* to the weaker sarcomeres. For convenience of comparison, curve 2. of Fig. 8.2 is repeated on a reduced scale below; it is the isometric contraction of the composite fibre.

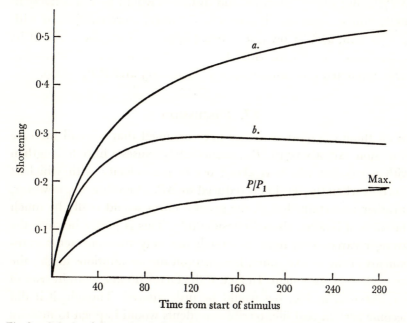

Fig. 8.3. Calculated shortening during isometric contraction 2. of Fig. 8.2, *a.* of the stronger sarcomeres and *b.* of the weaker sarcomeres. Below, marked P/P_1, is a reduced replica of curve 2., Fig. 8.2, with its maximum indicated on the right.

After the comparatively small differences between curves 1. and 2. in each of Figs. 8.1 and 8.2, the large difference shown in Fig. 8.3 was unexpected. But one can see what is happening; the shortening of the weaker sarcomeres ends at time 140 units, and then a very slow lengthening sets in; the shortening of the stronger sarcomeres continues for a long time, indeed indefinitely, and the maximum tension (Fig. 8.2) of the composite muscle is not reached by 400 time units. The rate of shortening of the stronger sarcomeres becomes very small, and could continue for seconds without the sarcomeres getting to a length at which their tension would fall off. But this is for a choice of P_2 only 20 % less than P_1; with a bigger difference much greater changes might occur.

§6. Units

No absolute units are attached to the figures. Tension is given only as P/P_0. But times, velocities and distances would be respectively in msec, mm/sec and mm if a muscle (or fibre) were considered with $l_0 = 31$ mm, $a/P_0 = 0.25$, $b = 0.325\ l_0/\text{sec} = 10$ mm/sec; and if its series elastic component, with a tension–extension curve of the normal form, stretched 0.8 mm between $P/P_0 = 0.05$ and $P/P_0 = 1.0$.

§7. Discussion

In all these calculations it has been assumed that a small increase of tension, ΔP, above its 'P_0', immediately causes a fibre to lengthen with a velocity which is about 0.2 of the velocity with which it shortens when the tension is reduced by ΔP. The value 0.2, however, is rather uncertain though the general effect would really be much the same if it were altered considerably. This is partly because the stronger sarcomeres have to stretch not only the contractile components of the weaker ones, but also (on the assumptions made) the series elastic component. In fact, if the factor became zero instead of 0.2, curve 2. in Fig. 8.2 would be little altered. Though, if it did become zero in real life, terrible accidents would happen to muscles unable to 'give' when impulsively stretched. In his paper (1957, p. 291) on *Theories of Contraction* A. F. Huxley considered the nature of the slow 'give' in a muscle subjected to a force greater than its P_0. From the constants he used in his equations he calculated the factor (taken above as 0.2) to be 0.23.

One interesting possible complication to the simplicity of the statement of the problem as hitherto set out was described in ch. 5, §2, particularly in connexion with Fig. 5.6 there. When a muscle was rapidly stretched early during an isometric tetanus, in order to obtain external evidence of its arrival at full activity, it invariably happened that the tension reached, and maintained after the stretch ended, was greater than what could be developed (at the same length) in an uncomplicated contraction. It could even happen that after such a stretch was over the tension continued to rise very slowly. Now this happened in a complete muscle after a rapid stretch; and I know of nothing to indicate whether anything similar does or does not occur in a single fibre. Does a slow stretch in a single fibre during full activity, such as that applied by stronger sarcomeres to weaker ones, raise the tension of the latter permanently during a continuing stimulus? or does the tension drop back as soon as the stretch stops? One can imagine a rapid stretch applied to a whole muscle causing a forcible and rapid redistribution of length between fibres, and putting them in a condition to hold a greater tension. In a single fibre a slow stretch may *not* cause an increment of tension which persists when the stretch stops. If it does, however, some of this chapter would have to be rewritten.

That stronger sarcomeres in a single fibre do actually stretch weaker ones was shown by Gordon, Huxley & Julian (1967). But their observation was made under the abnormal condition of a fibre at a length more than 50 % greater than l_0. When the fibre was stimulated the sarcomeres near the tendons stretched those in the middle of the fibre. This could be due to the sarcomeres near the tendons having the support of a stronger parallel elastic component than those elsewhere, not to a greater intrinsic strength. This stretched component would add to the tension developed by the contractile elements inside it.

When making force–velocity curves by the usual isotonic method I have sometimes noticed that the curve comes down to the base-line (at $v = 0$) more slowly than usual. The normally accepted curve meets the base-line at a finite angle at $P = P_0$, namely, $-dv/d(P/P_0) = b/(1 + a/P_0)$, which is quite a substantial number. The abnormality mentioned does not generally occur, the equation is usually obeyed with fair accuracy at tensions not far short of P_0. But Fig. 8.1 shows an evident difference between curves 1. and 2. in this region.

Again, I have sometimes noticed an isometric myogram creeping unusually slowly to its maximum. This was not due to 'fatigue', or to too high a frequency of stimulation, since the tension could be held quite well after the slow approach to its final level was over.

Fig. 8.2 illustrates how a 20 % difference of P_0 between two otherwise equal groups of sarcomeres could have this effect.

One final point of possible interest. It has been assumed throughout these calculations that the velocity constant b of the force–velocity relation is reduced to one-fifth when shortening is changed to stretch. What would happen if b were the same during lengthening and shortening? The calculation was repeated for the same two sets of sarcomeres as before, one set with a maximum tension 80 % of the other's: but with the same value of b in each direction, shortening or stretch. The tension of the stronger sarcomeres settled to 11 % less than P_1 while stretching the weaker ones, instead of the 3·5 % when the factor was 0·2; and the velocity of shortening of the stronger sarcomeres during a nearly steady state of stretching the weaker ones was increased threefold. The result was not as absurd as I expected. Perhaps in certain circumstances this might even be a desirable characteristic; though not in the Olympic Games.

References

Gordon, A. M., Huxley, A. F. & Julian, F. J. (1967). *J. Physiol.* **184**, 143.
Huxley, A. F. (1957). *Prog. Biophys.* **7**, 257.
Katz, B. (1939). *J. Physiol.* **96**, 45.

9

ODDS AND ENDS

A few odd facts, calculations or speculations were left over after most of the material had been tidied up or abandoned. Perhaps one or two of them may still interest somebody.

§1. The effect of a stretch on a collapsing structure

This seems rather an obscure title; its purpose is to induce someone to try to make the nature of the effect less obscure.

A. *A quick exponential stretch*

A stretch applied during a twitch, at a time when the active state has begun to pass off, causes—after an immediate rise of tension—a more rapid decay of tension than would otherwise have occurred. This is illustrated in Figs. 9.1 and 9.2.

The superimposed records of tension in Fig. 9.1 show the effect of a short quick exponential stretch applied at various moments during an isometric twitch of a toad sartorius at o °C. Line 1. is the record of the twitch running through three sections. Stretches 2. and 3. were applied before the active state had begun to decay and their tension records continued above the isometric level. In all the rest the tension soon dropped below it. In every case (including 2. and 3.) the muscle 'gave', as shown by the fall of the tension curve after its maximum while the stretch continued. In 2. and 3., 1 mm applied stretch was too much for a muscle which had already completed a good deal of its internal shortening (with stretching of its series elastic component); so it had to 'give', though it then 'recovered' and continued above the isometric level. In 4.–7. there was no recovery, the tension fell below what it would have been without a stretch. The arrows give the moments when the stretches were 95 % ($= 1 - e^{-3}$) complete.

Fig. 9.2 gives a similar series on a frog sartorius. In the early stretches, 2., 3. and 4., there is no sign of the tension dropping below

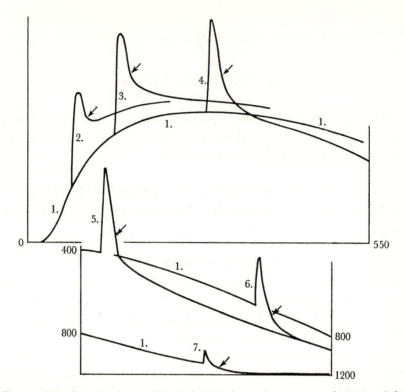

Fig. 9.1. The effect of a short quick stretch applied at various moments during a twitch. Toad sartorius, 0 °C, $l_0 = 32$ mm, stretched from $l_0 - 2 \cdot 5$ mm to $l_0 - 1 \cdot 5$ mm. Curve 1. is the isometric myogram, maximum tension 85 g, at length $l_0 - 2 \cdot 5$ mm; 2.–7. are the changes produced by an exponential stretch with a time constant of $7 \cdot 4$ msec. The curve is shown in three segments; a., 0–550 msec; b., 400–800 msec; c., 800–1,200 msec. The arrows give the moments when the stretches were 95 % complete.

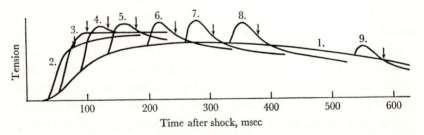

Fig. 9.2. The effect of a quick stretch applied to a muscle at various moments during a twitch. Frog sartorius, 0 °C, stretched from $l_0 - 0 \cdot 5$ mm to $l_0 + 1 \cdot 5$ mm. Curve 1. is the isometric myogram; 2–9 are the changes produced by an exponential stretch with time constant $13 \cdot 2$ msec. The arrows show the moments when the stretches were 95 % complete.

the isometric curve. They were made before the active state would be expected to have started to decay. No. 5. is betwixt and between, while 6., 7., 8. and 9. all fall clearly below.

B. *A slow linear stretch during a twitch*

This is illustrated in Fig. 9.3 for a frog sartorius. A stretch of 4 mm started at 17 msec and ended at 309 msec. The tension rose far above the isometric level and then gradually fell towards it as the stretch continued. When the stretch ended the tension dropped rapidly to far below the isometric curve. At the moment when the stretch finished the length of the muscle was the same in both contractions.

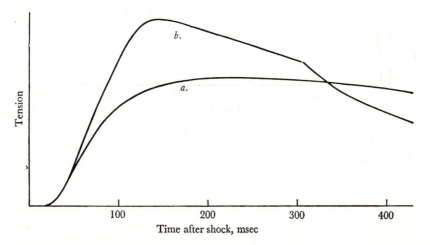

Fig. 9.3. Comparison of myograms, *a.* of an isometric twitch, and *b.* of an otherwise similar contraction during and after a slow linear stretch of 4 mm to the same final length ($l_0 - 2$ mm to $l_0 + 2$ mm, start 17 msec, end 309 msec). Maximum tension, isometric 64 g, stretch 94 g. Frog sartorius, 0 °C.

C. *A slow exponential stretch during a twitch*

This is illustrated in Fig. 9.4. A. was an ordinary isometric twitch of a frog sartorius, B. a twitch during which, from 18 msec onwards, the muscle was stretched at a decreasing rate. Curve B. crossed A. at 430 msec, then it remained below A. to the end of the contraction, in spite of the stretch continuing (at a decreasing rate).

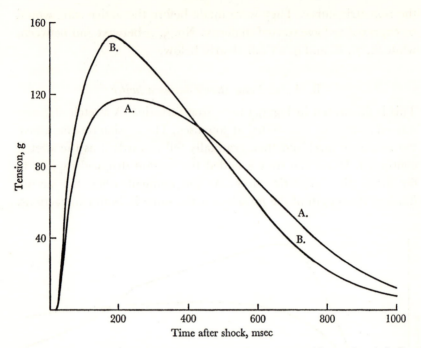

Fig. 9.4. Two contractions of a frog sartorius at 0 °C. A., a normal isometric twitch; B., a twitch during which the muscle was subjected to a slow exponential stretch of 2·1 mm with time constant 0·225 sec. B. and A. were successive contractions.

D. *Discussion*

In all these figures, while the active state was decaying, the muscles apparently showed a *negative resistance to stretch*—whatever that means. It is perhaps the sort of thing that an engineer would expect in a collapsing structure; it occurs consistently. A. F. Huxley's theory (1957) might explain it in less intuitive terms.

§2. A peculiar effect observed during relaxation after tetanic contractions of toad sartorii

In eight experiments, with isometric tetanic contractions of toad sartorii made over a period of two months, without exception the following peculiar effect was observed. After a normal tension rise and maintenance of maximum tension, when the stimulus ended the tension fell in what seemed to be normal relaxation: but not to zero, only to an average value of about $0 \cdot 1\ P_0$ (range $0 \cdot 03 - 0 \cdot 19$). At this level it remained steady for quite a time; then it dropped back slowly to its original level. See Fig. 9.5 for a contraction at 14 °C.

The toads (*Bufo bufo*) were of a variety described as 'continental'. They are referred to in ch. 5, §§1 and 3, and another record at 0 °C is in Fig. 7.8.

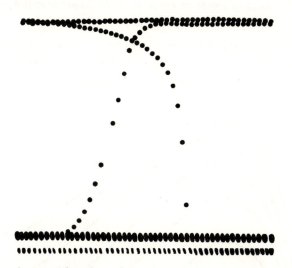

Fig. 9.5. Isometric contraction of a toad sartorius at 14 °C. 2 sec stimulus, 50 shocks/sec, alternating in direction. Ten 1 sec sweeps, spots every 0·02 sec. Maximum tension 129 g ($Pl_0/M = 3900$), remaining tension after apparently normal relaxation 6 g. Read from right to left.

The animals were in excellent condition and were being used for the many experiments described in ch. 5, §3. They were kept, as usual, at a low temperature. The effect was never observed after single twitches, or in the muscles of frogs kept under similar conditions.

I have had no opportunity of pursuing this curious effect further. Had the muscles been on a thermopile it would have been possible to decide whether the prolonged tension was accompanied by metabolic activity; caused perhaps by delay in pumping Ca^{2+} back to the vesicles. But the effect could possibly be due to these muscles having a substantial proportion of very slow fibres, taking a long time to complete their relaxation; though the appearance of the records suggests something more abrupt than that. The absence of any sign of this effect in twitches suggests that these slow fibres, if they exist, cannot be excited by a single shock.

§3. The effect of muscle length on the form of the isometric myogram of the sartorius, in twitch and tetanus

A. *The effect of length on the form of the isometric twitch, in the region of low initial tensions*

In Fig. 9.6 are copies of the records of five isometric twitches of toad sartorii at 9 °C, at lengths from $l_0 - 4$ mm to $l_0 + 0.8$ mm. The records were taken in succession from below upwards, at intervals of 2 or 3

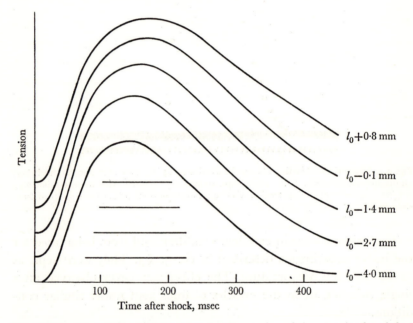

$l_0 + 0.8$ mm

$l_0 - 0.1$ mm

$l_0 - 1.4$ mm

$l_0 - 2.7$ mm

$l_0 - 4.0$ mm

Tension

100   200   300   400

Time after shock, msec

Fig. 9.6. Isometric twitches of toad sartorii at 9 °C, records made in succession from below upwards, at lengths shown on the right. Maximum tensions in order: 59 g, 68 g, 70·5 g, 72 g, 68·5 g. Base-lines for the last four are given by the horizontal bars.

min. For all of them the initial tension was quite small. The maximum developed tension (72 g) was at length $l_0 - 0.1$ mm. It was 59 g at length $l_0 - 4$ mm; the tetanic tension at that length was 117 g ($Pl_0/M = 2{,}620$).

Gordon, Huxley & Julian (1966*b*) investigated the effect of length on the maximum tension developed in tetanic contractions of single muscle fibres of frogs. The tension was greatest at a striation spacing of $2 \cdot 13 \, \mu$ and fell to 97 % with a 7 % reduction of length. With a greater reduction of length the tension fell relatively more. But it would be precarious to argue from frogs to toads, and from tetanic contractions to twitches.

The chief effect, however, of length in Fig. 9.6 was on the time course of the contraction. At greater initial lengths, even in the narrow range of the figure, the peak was later and relaxation more delayed.

B. *The effect of length on the form of an isometric tetanus*

In Fig. 9.7, in each of four experiments, records of tension are compared at two widely different lengths. The longer time to maximum at greater length is partly due, no doubt, to the same influence as produced the change illustrated in Fig. 9.6, where the length differences were smaller. But it is largely caused by the instability at greater lengths now to be considered.

In a muscle contracting isometrically during tetanic stimulation at ordinary length (not greater than l_0), if one element, sarcomere or region, is intrinsically stronger than another, the stronger tends to stretch the weaker. But the system is stabilized by the fact that when one element shortens its tension falls, in the other element, while it is being stretched, the tension rises. The tension is bound to remain the same in each since they are in series; but a slow exchange of length occurs between them, as discussed in ch. 8. At greater muscle lengths, however, this does not happen and the system becomes unstable.

The curve in Fig. 9.8 gives the *tension developed* by stimulation at constant length as a function of length. Of course at lengths sensibly greater than l_0 a muscle is under a standing tension before stimulation, but this tension, as is shown in §4 below, is not taken by the contractile filaments, but by extraneous structures. It is the filaments in which developed tension varies with length, so Fig. 9.8 is drawn in terms of developed tension.

At a length less than that for maximum developed tension, consider two elements in series at point A (Fig. 9.8). They are at the

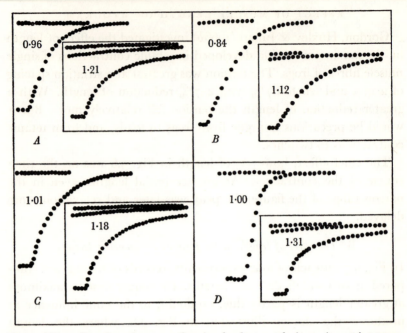

Fig. 9.7. Effect of length on the course of tension development during an isometric tetanus. *A.*, *B.*, *C.*, toad sartorii; *D.*, frog sartorii, all at o °C. In each pair the left upper record was made at or below the length for which the maximum tension developed was greatest, the right lower record at well beyond that length; lengths are given on the records as multiples of l_0. Repetitive sweeps in *A.* and *C.* are about 1·4 sec, in *B.* and *D.* about 1·0 sec. The records were made with 50-cycle spots; after the first five spots (after 10 in *D.*) alternate spots are omitted (Hill, 1953).

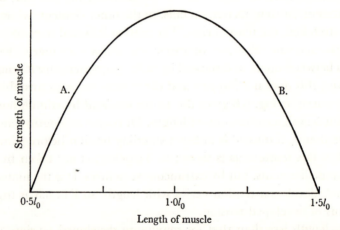

Fig. 9.8. Diagram to illustrate how the 'strength' of a muscle varies with the length at which it is stimulated. The 'strength', at any length, is taken as the greatest force it can develop at that length.

10

THE TRADITION OF THE INTERNATIONAL CONGRESS OF PHYSIOLOGISTS AND ITS FULFILMENT IN JAPAN

(Inaugural address to the Congress of the International Union of Physiological Sciences, Tokyo, 2 September 1965.)

This address is not directly related to the title of the book. But Doctor Ryotaro Azuma, who was the principal cause of my going to Japan in 1965, was working in my laboratory in London at the time of the First Experiments with Herbert Gasser in 1923–24; and it was during an interval taken from the Last Experiments that the address was given in Tokyo.

Seventy-six years ago, in 1889, the first International Congress of Physiologists met in Basel, Switzerland. Michael Foster, who was one of the chief agents in calling it, was very concerned that the meeting should be friendly and informal; so at the inaugural session he planted himself in the front row and rather ostentatiously smoked his pipe throughout. The French physiologist Louis Lapicque, on the other hand, then a young and naïve assistant (those are his own words), did not consider that it would be fitting for him to approach so many famous physiologists without a certain ceremony; so he appeared in a frock coat and a tall top hat. This garb gained for him, as he recorded nearly fifty years later, a pleasing and hilarious success. Thus was established, and continues, the tradition of friendly informality; it is characteristic of gatherings of physiologists everywhere. In recognition of his services to international physiology, and as a sign of the regard and affection of his colleagues, Foster himself, at the fifth Congress at Turin in Italy in 1901, was formally appointed Honorary Perpetual President of the Congresses. He died in 1907; but I like to fancy him here today, sitting below in the front row, smoking his pipe. His influence at least still lives.

The first Congress I attended, the ninth, was at Groningen in the Netherlands in 1913. There were now about 430 members, compared with 130 in 1889 and 3,000 today. It confirmed the tradition that an atmosphere of simplicity is natural and wholesome for

same tension, so the stronger s. is tending to shorten, the weaker w. to be stretched. As this slow process goes on the strength of s., moving down the curve to the left, diminishes; but the strength of w., moving up the curve to the right, increases. This stops the process from going any further. The system is stable.

Now consider what happens at B. on the opposite arm. The stronger and the weaker element are, and must remain, at the same tension. But the stronger element s. by shortening moves up the curve to the left and becomes still stronger; while the weaker w. by being stretched moves down the curve to the right and becomes still weaker. The system is inherently unstable. The stronger element continues to shorten and the tension rises; the weaker element continues increasingly to be stretched.

This explains the obvious differences between the pairs of contractions shown in each panel of Fig. 9.7: details of the four experiments are in the legend. The effect was recognized by Gordon, Huxley & Julian (1966b, p. 176) as the cause of the slow creep they observed in isometric contractions of single fibres at great lengths.

§4. The resting tension in a frog sartorius at lengths above l_0

The resting tension of an intact muscle at great lengths, as distinguished from that of a single fibre, is much larger than at l_0 where it is quite small. Measurements on single fibres at great lengths do not allow much of the extra tension in intact muscles to be attributed to the sarcolemma. According to Ramsey & Street (1940), working with single fibres at about 1·5 times the length for maximum tension development (P_0), the resting tension was not greater than 2 or 3 % of P_0. Gordon, Huxley & Julian (1966a) found rather less; at a striation spacing of 3·26 μ, which also is about 1·5 times the length for maximum tension development, the resting tension was only 1–2 % of the developed tension at optimum length (see their Fig. 13). And probably most of this resting tension resides in the sarcolemma anyhow, *so the tension in the resting filaments must be small up to any length practically attainable without damage in a whole muscle.* But even if *all* the observed resting tension were in the sarcolemmas of the fibres this could not possibly account for the tension in a stretched intact muscle; in a frog sartorius a 40% stretch from l_0 can produce a tension which is as great as the maximum developed tension at l_0 (see Table 1 below), and in the gastrocnemius the resistance to stretch is enormous. There must be elastic elements in a whole muscle far stiffer than the constituent fibres themselves.

In a frog sartorius at rest a good logarithmic relation is found between length and tension over a range of lengths from about 0·8 to 1·25 l_0. This relation cannot be of more than a statistical nature depending on the presence of passive structural elements, for nothing like it is found in single fibres. If P is the resting tension at length l and Q the tension at length l_0 the average relation (1949, Fig. 6) is:

$$\log P/Q = 4\cdot77\,(l/l_0 - 1).$$

Above 1·25 l_0 the relation ceases to be logarithmic, the tension becoming considerably greater than given by the equation (1952, Fig. 1). The transition from the simple logarithmic to the more complicated form starts when the elasticity begins to change over from 'thermokinetic' to normal. It is also at about this point that the 'stretch response' (Feng, 1932; Clinch, 1968) becomes obvious.

In a typical sartorius with a maximum developed tension, at length l_0, of 50 g and a resting tension at l_0 of 0·33 g, resting tensions at various other lengths are given in the second row of Table 1. They were calculated from the equation given above with $Q = 0\cdot33$. The last three entries in the second row are in square brackets because the formula does not apply above $l = 1\cdot25\,l_0$. The values printed below them were inferred from pp. 468–70 of my 1952 paper.

Table 1. *Effect of length on the resting tension and the maximum developed tension of the sartorius of* Rana temporaria

l/l_0	0·9	1·0	1·1	1·2	1·25	1·3	1·35	1·4	1·62
P, g (at rest)	0·11	0·33	1·0	3·0	5·2	[9·0] 10·6	[15·6] 24	[27] 63	—
Sarcomere length (μ)	2·02	2·25	2·48	2·70	2·81	2·92	3·04	3·15	3·65
P maximum, calculated, (g)	—	50	42	34	30	26	22	18	0

The last two tensions given under the second row are really very uncertain, because stretching a whole muscle to great lengths appears

to change it irreversibly; or at least to leave it in a condition from which it takes a long time to recover. They show the general effect, but quantitatively must be regarded with caution.

The third row of Table 1 gives the sarcomere lengths corresponding to the values of l/l_0 in the first row, assuming that at $l = l_0$ the sarcomere length is 2·25 μ. The last row gives values of the maximum tension developed ($P_{\max}$) in isometric contractions at the various lengths. The value at $l = l_0$ is assumed to be 50 g, the other values were calculated on the assumption that $P_{\max}$ falls linearly from its greatest value at 2·25 μ to zero at 3·65 μ, as Gordon, Huxley & Julian (1966 b) found.

These high values of the tension at rest, and the damage it seems to cause, make the whole muscle unsuitable for investigations at great sarcomere lengths. Whenever they can be used single fibres are better, as the work of Huxley and his colleagues has shown. But, for technical reasons, single fibres cannot always be used; for studying heat-production, for example, no instruments at present conceivable could be sensitive or quick enough. There are not many experiments, however, that seem to be worth making at great sarcomere lengths, and Huxley and his colleagues have made most of them already.

References

Clinch, N. F. (1968). *J. Physiol.* **196**, 397.

Feng, T. P. (1932). *J. Physiol.* **74**, 441.

Gordon, A. M., Huxley, A. F. & Julian, F. J. (1966 a). *J. Physiol.* **184**, 143.

Gordon, A. M., Huxley, A. F. & Julian, F. J. (1966 b). *J. Physiol.* **184**, 170.

Hill, A. V. (1949). *Proc. Roy. Soc.* B, **136**, 420.

Hill, A. V. (1952). *Proc. Roy. Soc.* B, **139**, 464.

Hill, A. V. (1953). *Proc. Roy. Soc.* B, **141**, 114.

Huxley, A. F. (1957). *Prog. Biophys.* **7**, 257.

Ramsey, R. W. & Street, S. F. (1940). *J. cell. comp. Physiol.* **15**, 11.

physiologists. Not that all was simple. After fifty-two years I still remember the banquet; the hospitality of Dutchmen is famous and there were about twenty courses, with speeches in between. One's digestion apparently, if not one's attention to speeches, reached a steady state. A year later the First World War intervened, and it was to be ten years before physiologists could have a fully international meeting again; this time, appropriately enough, under the presidency of Edward Sharpey Schafer, who had lost two sons in the war. But the same friendliness and informality remained. These were very evident once more at the Congress at Zürich in 1938 when W. R. Hess was president; but, again, general war was to prevent physiologists from meeting internationally for another nine years.

In 1947, however, at Oxford, under the genial presidency of Henry Dale, the old customs and traditions were revived. He attended his first Congress at Heidelberg in 1907, and read a paper on ergot; and he asked me recently to bring his affectionate greetings to his many friends here, particularly, as he said, to the Governor of Tokyo and his brother; both worked in Dale's Institute at Hampstead about forty years ago. May I suggest that we send to Dale[1] from this Congress our love and admiration? And since Hess is nine years senior to Dale in the presidency, though six years junior in years, may we send similar greetings to him?

This is the eleventh Congress I have attended and the membership has increased manyfold. The increase has provided serious problems for the organizers, with which, however, they have always been able to cope with efficiency and kindness. No doubt at the closing session thanks will be offered to our Japanese friends for all they have done for us; but I have been here already for a few days and have seen what a wonderful job Professor Kato and his colleagues, and the City of Tokyo, have accomplished. They do it with the utmost grace and efficiency—and make it clear to us all that we are their very welcome guests.

My lasting impression of these Congresses is of the supreme value of the personal contacts and friendships they have started or revived. Without these, all our laboratories everywhere would have been much poorer. I confess with shame that I have sometimes been

---

[1] Dale died in July 1968, at the age of 93.

rather a meagre attendant at the scientific sessions. More profitable to me, for the most part, have seemed the informal meetings and discussions outside; with the two-way traffic of facts, ideas and reminiscences, with new friends and old, as opportunity served. I know that this is a deplorable weakness, and it increases with the years; but it is one which I share with some others—fortunately for me, for they are the ones I have found outside.

During the early 1950s the physiologists of the world formed an International Union of Physiological Sciences, IUPS, which in 1955 was accepted as a Scientific Member of ICSU, the International Council of Scientific Unions. It is one now of a family of fifteen independent Scientific Unions, and of various other organizations such as the International Biological Programme. Before 1955 the International Congresses were the only joint meeting ground of world physiologists. These Congresses then had no legal existence at all, and a minimum of formality or structure. Custom, and the wholesome tradition started in 1889, and of course the professional readiness of physiologists to make experiments, were the only guide and bond. Somehow they got on very well for 64 years; but under the new régime there will be greater opportunities than of old for physiologists to meet and cooperate with each other, and with members of other unions of kindred scientific interests—not only during Congresses but at other times. The only fear might have been that some of the old tradition of informality would be lost. It has not been lost, as we shall find again at this Congress; the common sense of physiologists everywhere, and the understanding of our hosts, will see to that.

In the past these gatherings have been called Congresses of Physiologists, that is, of ourselves. Now they have become Congresses of IUPS. But we come here still as ourselves, as individual physiologists, representing nothing and nobody else. Let me tell you here a pertinent little story. Long ago there was a tremendous strike in the London Docks and something had to be done to keep supplies flowing. So great lorries were being driven in by soldiers, with formidable notices on them, 'By the authority of His Majesty's Government', by the authority of this or that. And among them was a tiny donkey cart, driven by a little old man in a broken bowler hat; and on his cart was a home-made notice 'By my own bloody authority'. Not all of you perhaps will understand the full derisive impact of that vulgar English idiom; but anyhow I am glad to think that physiologists still come to these Congresses in the independent spirit of that little old man in his donkey cart.

The motto of the Royal Society of London, *Nullius in verba*, is derived from a Roman poet who wrote: 'Not being bound by the authority of any master, wherever the wind blows me I come into port as a welcome guest.' Today the winds have blown us to a lovely land and a charming and hospitable people, and we look forward with the keenest pleasure to all we shall experience here. And to some of us who have been fortunate enough to work with Japanese colleagues, there is the supreme happiness now of meeting them here at home. Of these colleagues, the earliest to come and work with me was Ryotaro Azuma, who turned up in 1923. The first thing, it is said, that he did in London was to join the Thames Rowing Club; and at intervals, as he returned, he went off there again to revive happy memories and renew his youth. All his life he has been a practitioner, a pioneer and a prophet in the development of sport and athletics in Japan, as healthy, humane and civilizing pursuits; which also can act and react with physiology and medicine to their mutual benefit. To that I will return later on.

Another Japanese friend, Furusawa, spent six strenuous years with us and contributed greatly to the results obtained in several fields of research. So long indeed did he stay, and so English did he become, that—to avoid confusion as we said—we added a Scottish prefix to his name, and he still signs himself, at least to us, as Mac Furu. Another, Hukuda, spent two years with us; my happiest memory of him is of a lovely afternoon on the Devon coast, when his playful porpoise-like adroitness in the sea allowed him to take ample revenge on an Italian colleague, who had teased him in the Plymouth laboratory. Rodolfo Margaria may remember that too! And there were several others. Their devoted co-operation and affection made them delightful colleagues; and the memory of their gay companionship in research recalls a second saying of the same Roman poet which can be paraphrased thus: 'There is nothing to stop you from announcing your results with a laugh.'

One purpose of these Congresses is to introduce their younger members to some of the famous men of whom they have heard: to help them to realize that discoveries are made by ordinary people like themselves, not by distant legendary figures. For example, at the Congress in Leningrad in 1935 I remember well the vivid interest of a group of young Russian physiologists in a spontaneous discussion they had with Joseph Barcroft. To a sympathetic observer it seemed that, for these youngsters, some kind of magic was transforming legend to human reality. The same magic attended the presence of Ivan Pavlov, the president, filled with energy and fun in spite of his 86 years. Whenever and wherever he appeared his romantic and almost fabulous figure aroused long and enthusiastic applause. Of this indeed he often seemed rather impatient; one could see him, at some of the plenary sessions, shaking his fists repeatedly and uttering hard words until the disturbance was over. But I think he enjoyed it all the same. Many honours had come to him, but the one he probably liked best emerged spontaneously, many years before in 1912, from the students at Cambridge, when he received an honorary degree. They had heard in lectures about his famous work on digestion; so they went and got a large and life-like dog from a toy shop, and proceeded to furnish it with all the glass and rubber tubes, fistulae and stoppers that there was room for. Then as soon as his degree was conferred in the Senate-House, they let down this strange object by strings from the galleries into his arms. He carried it off very graciously, remarking later 'why, even the students know about my work'—as though that was the last thing one would expect. The dog, with its trappings, went home with him to Leningrad and remained in his study for many years. It is now in the Pavlov museum, the director of which, a few years ago, wrote to ask me for details of its origin. It is not often that students' jokes are so welcome; but seldom can one have been so perfectly adjusted to the humour of its recipient.

There are many other happy memories of these Congresses of which an outstanding one is of 1929, when several hundred physiologists from more than twenty nations travelled together in a slow ship (eleven days) from London to Boston. The long voyage gave a wonderful opportunity for making friends and exchanging scientific, and other, experience; and also for an international tug-of-war contest on deck, in which young and old alike took part. Dale pulled for England, I wasn't heavy enough and was elected an honorary Swede. The Congress itself, at Boston, was followed by a mass excursion to Woods Hole; and a wonderful day ended with what the overseas visitors found an unusual entertainment, a 'clam-bake' and a 'lobster feed' on the sea-shore. I remember finding rather a lot of sand in my clams, but that I gather is not customary.

I promised earlier to return to the relationship between physiology and athletics. I have just been attending here, as others have, a remarkable symposium on the ama, the Japanese diving woman who performs almost unbelievable feats daily in the ordinary course of the family business of collecting pearls, shell fish and sea-weeds. For many years the achievements of these women have excited the wonder and interest of Japanese physiologists and others. In a country where such a singular traditional calling exists, it is not surprising that physical training and athletics should have built up a mutually profitable connexion with medical science. I was astonished, however, recently to learn from Dr Azuma that in a minor way I had something to do with this. In 1925 I gave an address to the British Association on 'The physiological basis of athletic records', which aroused Dr Azuma's interest and (so he said) helped to start him off working for a closer link-up between sport, physiology and medicine. We have often heard of science advancing by a chance seed falling into prepared ground—in his case the ground had been prepared by rowing! Anyhow in 1927 a small study-group was formed in Tokyo for 'Sports Medicine', to co-operate with athletes and their organizations. This, as Azuma told me, was the humble beginning of what has become a large-scale organization. Starting like this, the 'Japanese Association for Promotion of Physical Fitness' has developed so far that last year, in conjunction with the Olympic Games, an 'International Congress of Sport Sciences' was held in Tokyo, with more than 1,000 participants. The chairman of the Committee which organized that Congress was Toshiro Azuma, brother of Ryotaro; the latter as a member of the International Olympic Committee had been responsible for bringing the Olympic Games to Tokyo. Of course all this is perfectly well known to Japanese physiologists, but not so well perhaps to the rest of you. To Japanese sportsmen, athletes and physiologists, and particularly to the Azuma brothers themselves, the Tokyo Olympics of 1964, and the associated International Congress of Sports Sciences, are a fitting crown to nearly forty years of thought, endeavour and education. While to the people of Japan in general, the magnificent success of these great enterprises must be a matter of enormous pride and happiness. Friendly co-operation and competition in science, sport and athletics, supply a much nobler and less expensive ideal than war.

To physiologists, the Olympic Games provide a supreme example of human physiological experiments. And now that athletes, their coaches and the public in general are beginning to think in physiological terms much new and exciting knowledge will come from them. People are already beginning to think hard about the Olympic Games to be held in 1968 in Mexico City, at a height of 2,300 metres. A complete novelty there is being introduced by the fact that the partial pressure of oxygen is only 76% of that at sea-level; nobody can foretell exactly what the effect of this will be, but I should guess that, however well acclimatized, the long-distance runners, swimmers and oarsmen will have to go rather slower. On the other hand, the density of the air is correspondingly less, so air resistance to a sprinter will be slightly lower and he may be able to go just a little faster. But a lot of other interesting and more complicated effects are likely to turn up, which will need physiologists and physicians to interpret them.

It is many years since I took any part myself in physiological work on severe muscular effort as involved in athletics. But I retain a very vivid memory of the accuracy and consistency with which the trained runner can carry out his task and repeat it again and again. All physiologists know how greatly they depend on the quality of their experimental animals, in order to obtain reliable results; and a co-operating and highly trained animal, as the human athlete can be, is an enormous asset to the experimenter. The experiments made around forty years ago, and the technical methods that were available then, look very crude compared with those of today. There are indeed great possibilities of new and original experiments; and here in Japan, particularly, there must be a wealth of young trained volunteers ready to lend themselves for the purpose. I wish I could take part in their new and exciting adventure. Good luck to them.

May I return once more to our Congresses. Talking of athletics is bound to make one think of records, so if you will forgive the frivolity I will refer finally to a world's record that was set up at the Congress in Moscow in 1935. At the concluding session George Barger had been appointed to thank our hosts for their magnificent hospitality. He proceeded to address them in nine languages, in alphabetical order: Dutch, English, French, German, Italian, Latin, Spanish, Swedish and of course Russian. This record will not be easy to beat, except perhaps by a professional linguist disguised as a physiologist. But it presents a challenge to the Officers and Council of IUPS. Looking at the list of them I should guess that between them they could muster at least fifteen languages, though that is not quite the same thing as Barger's solo effort. We shall need, anyhow, something really distinguished on 9 September in thanking our Japanese friends.

With that gay suggestion may I end by saying how honoured I have been by this opportunity of addressing you, by your kindness in listening and your tolerance of occasional frivolity. And I am deeply grateful to the Governor of Tokyo, my dear friend of many years Ryotaro Azuma,[1] for having brought me here as the guest of the Metropolitan Government of his great and hospitable city; of which, last Monday, he presented me with the golden key. By some magic, I find, this key has already unlocked the hearts of the people of Tokyo.

[1] He is now (1969) President of the Japanese Red Cross Society and an honorary life member of the International Olympic Committee.

INDEX

Abbott, B. C., 27
action potential, muscle and nerve fibres, 55
active state, produced by stimulus, xi, 5, 21
  26, 34
  duration and decay, 16, 26, 56–75
  time-course of intensity, 15, 58
ama, 135
athletics, 133–7
ATPase activity, effect on muscle speed, 55
Aubert, X. 26
Azuma, R., 130–7
Azuma, T., 135–6

Bárány, M., 55
Barcroft, J., 134
Barger, G., 137
Basel (Congress), 130
Benedict, F. G., 23
Best, C. H., frontispiece, 26
Blix, M., 87
Borelli, G. A., 87
Boston (Congress), 134
Brabazon, Lord, 85
British Association, 135
Buller, A. J., 55

calcium, in relaxation, 123
Cambridge, xi, 134
Caplan, S. E., 25
cardiac muscle, force-velocity relation, 26
Cathcart, E. P., 23
cathode ray tube, 1, 2
  records with, 30, 39, 44, 69, 100, 123
Cavagna, G. A., 67
Clinch, N. F., 128
Close, R., 24
collapsing structure, effect of stretch on, 119–22
compliance, extra added during contraction, 27, 36, 37
connecting wire, 73 n
contractile component, 86

Dale, H. H., 2, 131, 135
Darwin, Horace, 28
Davies, R. E., 83
delayed isometric contraction, 22, 35, 45–51, 57, 94, 108
Dusman, B., 67

Efficiency of contraction, variation with velocity of shortening, 4, 5, 23
Elastic component, see series elastic component
  parallel, 117, 127
elastic energy in contracting muscle, 76–81
ergometer, 29, 30, 61, 76–9

Feng, T. P., 128
Fenn, W. O., 2, 10 n, 23, 25
Fick, A., 87
Flexner, A., 1.
force–velocity relation, xii, xiv, 10 n, 23–41,
  52–4, 86, 88, 93, 111, 118
  with added compliance, 36, 37
  b-constant during lengthening, 110, 116,
    118
  constants of, xiv, 23, 27, 92, 100, 102
  during isometric contraction, 33–7
  effect of an error in a/P_0, 92
  equations, xiv, 23–26, 53
  force at constant velocity, 28–32, 55
  heat of shortening, 24
  'instantaneous' property of muscle, 34,
    37–41
  and irreversible thermodynamics, 25
  in non-uniform fibre 111
  at shorter and greater lengths, 27
  in single fibre, 27
  in single twitch, 26
  speed and efficiency 4, 5, 23
  speed and work, 77
  statistical nature of, 52–5
  various muscles, 24, 25, 26
  velocity with constant force, 27
Foster, Michael, xiii, 130
Furusawa, K., 133

Gasser, H. S., xi, xii, 1–4, 23, 57, 130
Gordon, A. M., 27, 108, 117, 125, 127,
  129
Groningen (Congress), 130

Hartree, W., 78
heat production, initial, 15, 51
  relaxation, 78, 83
  shortening, 23–5
Heidelberg (Congress), 131
Hess, W. R., 131

139